Molecular Recognition: Chemical and Biochemical Problems II

Molecular Recognition: Chemical and Biochemical Problems II

Edited by
Stanley M. Roberts
Department of Chemistry, University of Exeter

ROYAL
SOCIETY OF
CHEMISTRY

The Proceedings of the Second International Symposium, Chemical and Biochemical Problems in Molecular Recognition, organised by the Fine Chemicals and Medicinal Group of the Industrial Division of the Royal Society of Chemistry. University of Exeter, 6–10th April 1992

Special Publication No. 111

ISBN 0-85186-226-8

A catalogue record for this book is available from the British Library

Published by The Royal Society of Chemistry,
Thomas Graham House, Science Park, Cambridge CB4 4WF

Printed in England by Redwood Press Ltd, Melksham, Wiltshire

Preface

The articles in this book reflect the proceedings of the second international conference on Chemical and Biochemical Problems in Molecular Recognition held in Exeter (April 1992). The authors are amongst the world leaders in this area and detail many of the advances that have been made in the field.

Molecular recognition has been studied in detail through measurements in small molecules such as triethylamine and HX. Some of the recognition phenomena that are observed make contributions in the interactions of large molecules/small molecules and large molecules/large molecules. For example hydrogen bonding is important in non-covalent binding of very simple substances. It is also crucially important in ligand-receptor, substrate-enzyme and protein-DNA interactions as well as in protein folding. Yet the energy change on formation of such a simple bond is still a matter of debate. The orientation of the hydrogen donor and the hydrogen acceptor is also crucial in this particular interaction.

Not only is spectroscopy (*e.g.* rotational spectroscopy and NMR spectroscopy) important in this sort of work but so are theoretical calculations employing the most advanced computer technology. Obviously detailed understanding of receptor-ligand and enzyme-substrate binding is essential if *de novo* design of useful drug substances and agricultural aids is to become a reality.

Not surprisingly there has been a lot of recent interest in self-assembly processes leading to the formation of two-dimensional and three-dimensional networks. Behind much of this work lies sophisticated organic synthesis and many of the recent innovations in methodology are incorporated into the preparative work.

While impressive progress has been made over the past few years in the area of molecular recognition it is also obvious that there are numerous intriguing problems that remain to be solved.

Contents

A Contribution to Molecular Recognition from Rotational Spectroscopy

A. C. Legon

DEPARTMENT OF CHEMISTRY, UNIVERSITY OF EXETER, EXETER EX4 4QD, UK

1   INTRODUCTION

At the fundamental level, molecular recognition involves the specific interaction of one part of a molecule with a particular part of another molecule. This interaction will be defined by a relative orientation and by a separation of the two subunits that confer on the system as a whole a lower energy than other conformations. An understanding of the fundamentals of molecular recognition therefore requires a knowledge of the properties of intermolecular interactions and in particular how the energy varies with relative orientation and separation. Biologically significant molecules are often (by chemical standards) rather large and exhibit a range of different sites, with each site capable of sustaining an energetically favourable interaction with a particular type of site or group in another molecule. Some sites will be more or less nucleophilic while others will be more or less electrophilic. For the purpose of modelling large systems, it is of interest to follow an approach familiar in chemistry, *i.e.* to consider the larger system to be composed of groups, each group having its own characteristic properties. Thus, we might enquire into the preferred angular and radial geometry for the interaction of one group with another and, given that in a larger unit it might not be possible to achieve the preferred arrangement, we might then ask about the cost in energy of small angular and radial distortions from this conformation.

In this chapter, we present some experimentally determined properties of pairs of small molecules allowed to interact in isolation. The interaction is usually between a model molecule B and a very simple probe molecule, such as HF, HCl, and so on. The advantage of small diatomic, dipolar molecules HX is that, as we shall see, the positive end $H^{\delta+}$ seeks out regions of high nucleophilicity on B in forming a hydrogen-bonded dimer B$\cdots$HX. By investigating carefully selected series of dimers B$\cdots$HX in which first B and then HX is systematically varied, it has been possible to make some generalisations about the properties of the interaction. From these generalisations have followed models that allow the nature of the interaction to be understood.

A powerful method of characterising hydrogen-bonded species B$\cdots$HX is rotational spectroscopy. It leads to a number of precise properties of B⋯HX and, because it is conducted at low pressure in the gas phase, these properties pertain to the unconstrained dimer in isolation. Two methods have been employed in the work described here: Stark-modulation microwave spectroscopy and pulsed-nozzle Fourier-transform microwave spectroscopy. The first technique relies on the absorption of microwave radiation by the species B⋯HX in an equilibrium mixture of B,HX and B⋯HX and is suitable for studying only strongly bound species such as H_2O⋯HF and HCN⋯HF, for which the equilibrium concentrations are detectable at accessible temperatures. On the other hand, the second technique involves observing rotational spectra of species B⋯HX formed by supersonically expanding a pulse of a mixture of B and HX diluted in argon through a nozzle into a vacuum. The pulse so formed is rich in species B⋯HX which are travelling in collisionless expansion when they interact with the radiation. The effective temperature of the pulse is then ~ 1K and even the most weakly bound dimers B⋯HX can be characterised in this way. Details of the two techniques have been reviewed elsewhere.[1,2]

2   WHAT PROPERTIES OF ISOLATED DIMERS B⋯HX ARE AVAILABLE FROM ROTATIONAL SPECTROSCOPY?

The detailed properties of isolated dimers B⋯HX that can be determined from rotational spectroscopy have been discussed[3] for the prototype dimer HCN⋯HF. Table 1 summarizes the spectroscopic observables and the dimer properties that can be established from them.

Table 1   Spectroscopic and molecular properties of B⋯HX available from rotational spectroscopy.

Spectroscopic quantity/effect	Molecular property
1. Type of spectrum	Symmetry of B⋯HX
2. Moments of inertia	Distribution of mass; radial and angular geometry.
3. Centrifugal distortion	Restoring force constants of the intermolecular bond.
4. Transition intensity (a) absolute intensity.	Dissociation energy D_0 for B⋯HX.
(b) relative intensity of a given rotational transition in different vibrational states.	Vibrational separations, especially in low energy, intermolecular modes.

Table 1 (continued)

Spectroscopic quantity/effect	Molecular property
5. Nuclear hyperfine effects	
(a) from electric quadrupole coupling of **I** and **J**.	Electric charge distributions at quadrupolar nuclei.
(b) from magnetic H,X spin-spin coupling.	Lengthening of HX bond on formation of B⋯HX.
6. Effects of applied electric fields on rotational transitions.	Electric dipole moment of B⋯HX and thence extent of polarisation of one subunit by the other.

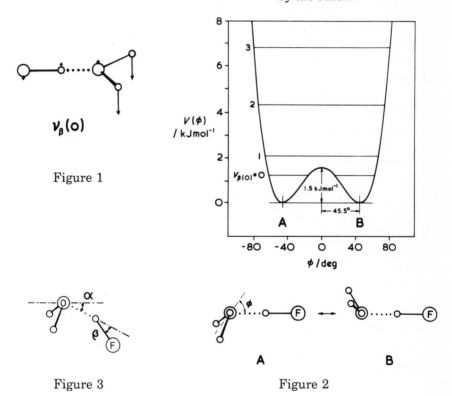

$\nu_\beta(0)$

Figure 1

Figure 3

Figure 2

3. $H_2O\cdots HF$ AS A PROTOTYPE DIMER

The dimer $H_2O\cdots HF$ has been extensively investigated[4-10] through its rotational spectrum by both methods mentioned in Section 1. All of the spectroscopic and dimer properties set out in Table 1 have thereby been determined and the molecule has been characterised in some detail. The main conclusions can be summarized as follows.

The nature of the rotational spectrum (including the presence of nuclear spin statistical weight effects like those familiar in H_2), the magnitudes of the observed moments of inertia and their changes on isotopic substitution, the vibrational separations from relative intensity measurements (especially $v_{\beta(o)} = 1 \leftarrow 0$ and $2 \leftarrow 0$ for the out-of-plane

bending mode $v_{\beta(o)}$ shown schematically in Figure 1), and the variation of the moments of inertia and the electric dipole moment with $v_{\beta(o)}$ lead unambiguously to the conclusion that $H_2O\cdots HF$ has an equilibrium geometry of C_s symmetry of the type indicated in Figure 2. Thus the HF molecule acts as proton donor in forming a hydrogen bond to oxygen. Moreover, the spectroscopic constants allow the quantitative variation of the potential energy $V(\phi)$ with the out-of-plane angle ϕ to be deduced[5], with the result shown in Figure 2. The essential conclusion is that although the molecule is effectively planar because the barrier to the planar form $\phi = 0$ is lower than the zero-point level (*i.e.* the vibrational wavefunctions have C_{2v} symmetry), the *equilibrium* conformation at oxygen is definitely trigonal pyramidal, as shown.

The various measures of the strength of interaction are also available. Absolute intensity measurements in the rotational spectrum give $D_0 = 34.3$ kJ mol[-1] for the energy required to separate the zero-point molecule into its components.[10] On the other hand, the force required to displace the zero-point molecule by unit infinitesimal distance along the dissociation coordinate has been determined to be $k_\sigma = 24.9$ Nm[-1] from centrifugal distortion effects.[11] The small extension $\delta r = 0.015$Å of the HF bond when the dimer is formed from its components has been evaluated from various nuclear hyperfine coupling constants and establishes that the simple hydrogen-bond model $H_2O\cdots HF$ is appropriate.[7,12] Contributions from the ion-pair form $H_3O^+\cdots F^-$ to the description are clearly negligible. Finally, the resistance of the hydrogen bond to bending in two ways has been assessed. The quadratic force constants $k_{\alpha\alpha} = 2.52 \times 10^{-20}$ J rad[-2] and $k_{\beta\beta} = 19.70 \times 10^{-20}$ J rad[-2] for bending the hydrogen bond through the angles α and β defined in Figure 3 have thus been determined.[8] It should be noted that the hydrogen bond is considerably more resistant to bending at hydrogen than at oxygen. When the angular distortion of hydrogen bonds is being discussed, this fact (which applies to other systems, including $HCN\cdots HF$) should be taken into account.

4   A SIMPLE MODEL FOR ANGULAR GEOMETRIES IN $H_2O\cdots HF$ AND OTHER $B\cdots HX$ DIMERS

It is obvious that a potential energy function of the type established experimentally for $H_2O\cdots HF$ can be explained if it is assumed that in the equilibrium conformation of the dimer the HF molecule lies along the axis of a conventional nonbonding electron pair on oxygen, with $H^{\delta+}$ of

HF sampling this nucleophilic centre. Because the interaction is relatively weak and H_2O carries two equivalent n-pairs in the conventional representation (see Figure 4), the observed low potential energy barrier to the planar form and the consequent ease of tunnelling between the two equivalent pyramidal conformers are readily understood.

Figure 4

In fact, a wide range of dimers B···HX has been investigated by rotational spectroscopy (although, except for HCN···HF, not as fully as H_2O···HF) and it has been possible to generalise the above interpretation into a rule[13,14]: *in the equilibrium conformation of a hydrogen-bonded dimer B···HX, the HX molecule lies along the axis of a nonbonding electron pair on B, as conventionally pictured.* Figure 5 shows schematic representations of the angular geometries of HCN···HF, H_2S···HF and H_2CO···HF expected on the basis of this rule. The observed angles ϕ are[1,15,16] 180°, 89° and 110° degrees, respectively, as expected from these familiar n-pair models. For more detailed discussion of the above rule in relation to observed angular geometries, see ref. 14.

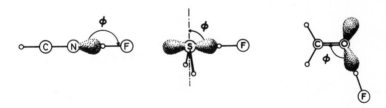

Figure 5

5   AN ELECTROSTATIC INTERPRETATION OF THE MODEL

Evidently, the simple rule discussed in Section 4 is electrostatic in origin. The electrophilic $H^{\delta+}$ of HF seeks the most nucleophilic region (*i.e.* the axis) of the n-pair on O. There is some evidence that the interaction between H_2O and HF (and between more weakly interacting B and HX species) does not lead to large geometrical and electrical perturbations of the components. Moreover, it has been shown[17] that any charge transfer and polarization contributions to the energy are not strongly dependent on the relative orientation of the subunits.

The question of the existence of n-pairs has been reviewed recently.[14] Briefly, the argument in support is as follows. If the interaction of B and, say, HF is electrostatic, the predominant contribution to the energy arises from the charge $\delta+$ on H and an n-pair on O. In this crudest approximation, therefore, the direction along which the nonperturbing point charge $\delta+$ has lowest energy corresponds to the direction of an n-pair *i.e.* the direction of minimum electrostatic potential delineates the n-pair direction. The variation of the electrostatic P.E. of a nonperturbing point positive charge near to a molecule B is nowadays readily available from *ab initio* calculations. The result $V(\theta)$ when a charge of $+e$ (*i.e.* a proton) is carried around an oxygen atom of SO_2 in the molecular plane and at the fixed distance $r(O\cdots H)= 1.89$ Å from O is shown in Figure 6. The distance corresponds to the experimental distance $r(O\cdots H)$ found in $SO_2\cdots HF$. We note that there are two potential energy minima, one at $\theta\sim240°$ and the other (slightly higher in energy) at $\theta\sim140°$. The first of these presumably corresponds to the direction of the n-pair that is *cis* to the S=O double bond (as illustrated in Figure 6) while the other corresponds to the *trans* n-pair. It is interesting to record that the lowest energy form of $SO_2\cdots HF$ (as observed through its rotational spectrum by the pulsed-nozzle, F-T method) has $\theta=215°$ and approximates to the *cis* isomer predicted by the rule.[18] Similar agreement between the directions of n-pairs and the observed angular geometries of B$\cdots$HX occurs for a wide range of B and HX.

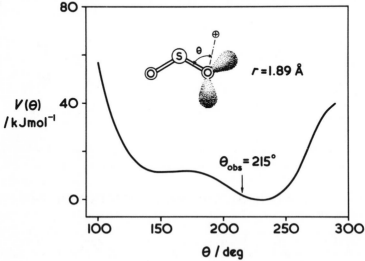

Figure 6

It turns out that the electric charge distribution of HF can be described with useful accuracy as a simple extended electric dipole moment $H^{\delta+}-F^{\delta-}$, where $\delta = 0.54e$. This allows a convenient description of the way in which the potential energy of molecules B and HF varies with relative orientation at the next level of approximation. When the above calculation is repeated with $H^{\delta+}-F^{\delta-}$ as the extended dipole instead of just the charge $\delta+$, the minima in electrostatic potential energy now occur at 240° and 120° *i.e.* along exactly those directions associated with the n-pair model.

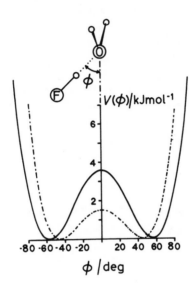

Figure 7: ····· Observed potential function $V(\phi)$
————— Calculated function (see text)

The result for $V(\phi)$ when HF is confined to the plane perpendicular to the molecular plane of H_2O and the linear arrangement O···H-F is maintained with $r(O\cdots H)$ fixed at the experimental distance in $H_2O\cdots HF$ is shown in Figure 7. Also included in Figure 7 is the experimental potential energy function $V(\phi)$. Given the crudeness of the model, the agreement between the two functions is remarkable. Note that $\phi_{min} = \pm 55°$ again delineates the directions conventionally associated with n-pairs in H_2O. A successful and complete electrostatic model in which the charge distributions of both B and HX are described by a distributed multipole analysis gives good agreement with observed angular geometries.[19]

6   SOME PREDICTIONS BASED ON THE n-PAIR MODEL

If the n-pair model is a good one, it should be possible to vary the molecules B and HX in a systematic manner and predict the accompanying changes in the angular geometry of B···HX on the basis of the model. We start from H_2O···HF.

The first variation is to replace the H atoms in H_2O by alkyl groups. The accepted view is that alkyl groups are electron releasing relative to H. If so, the n-pairs on O should become more nucleophilic and the O···H-F interaction should increase in strength. Presumably, the potential energy barrier at the planar ($\phi=0$) form previously alluded to in H_2O···HF should increase. A convenient model molecule here is 2,5 dihydrofuran, for the COC angle is approximately tetrahedral (114.4°) but the rigid pseudo four-membered ring is spectroscopically more convenient than *e.g.* $(CH_3)_2O$ would be. The observed angular geometry[20] of the 2,5 dihydrofuran···HF complex is illustrated in Figure 8. No evidence of inversion is apparent in the rotational spectrum of this dimer and the configuration at O is now permanently pyramidal, as anticipated. The angle ϕ = 48.5° is again close to half the tetrahedral angle, as predicted if HF lies along the axis of one of the tetrahedrally disposed n-pairs carried by O.

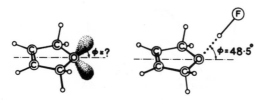

Figure 8

The second variation is to close up the angle $\hat{COC}$. It is generally accepted that as the $\hat{COC}$ angle decreases its *p*-character increases and the *s*-character of the n-pairs should therefore increase commensurately. The $\hat{COC}$ angle can be decreased effectively by replacing 2,5 dihydrofuran first by oxetane and then by oxirane. The observed angular geometries[21-22] of B···HF and conventional n-pair models of B are given in Figure 9. The expected increase of ϕ from B= 2,5 dihydrofuran, through oxetane, to oxirane is clearly observed. Recent observations for various other complexes oxirane···HX and thiirane···HX are also in agreement with these conclusions.[23-25]

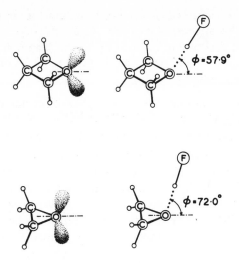

Figure 9

The third variation is to leave B fixed but vary HX in such a manner that the hydrogen-bond interaction becomes progressively weaker. If the HX molecule does indeed form a hydrogen bond by interacting electrostatically with an n-pair on, say, an O atom, the weaker the interaction the less directing will the effect of the n-pairs be. The simple electrostatic model used in Section 5 to predict the potential energy function $V(\phi)$ in $H_2O\cdots HF$ (Figure 7) gives support to this intuitive assumption.[26] Figure 10 shows the result when the simple

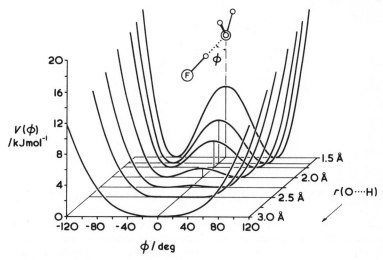

Figure 10

extended dipole model of HF is used to calculate $V(\phi)$ exactly as described
in Section 5 but for a range of $r(O \cdots H)$. At short range a relatively deep
double minimum results but the barrier to the planar form ($\phi=0$) drops
off rapidly and even for distances as small as $r = 2.5\text{Å}$ only a flat single
minimum is discernable.

An experimental manifestation of this effect has been obtained for
the series $H_2CO \cdots HX$, where X=F, CN and CCH. While $H_2CO \cdots HF$ is
rigid with no inversion and the angle $\phi=110°$ (see Figure 11)[16], evidence
for inversion has been adduced in $H_2CO \cdots HCN$ and the angle $\phi = 138°$
approaches the limit of 180° expected if the n-pairs on O have no
directing effect.[27] The hydrogen bond interaction in $H_2CO \cdots HCCH$, on
the other hand, is so weak that a secondary interaction (between either
$C^{\delta+}$ or $H^{\delta+}$ of H_2CO or both and the $C \equiv C$ triple bond) becomes
important.[28] The cyclic structure shown in Figure 11 is unexpectedly
rigid.

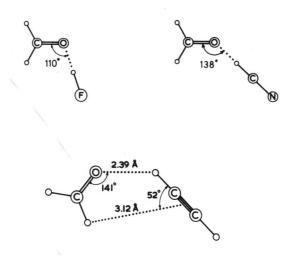

Figure 11

7   B AND HX AS AMPHIPHILES: ISOMERS OF B···HX

We have seen that the interaction of B and HX in the cases considered
can be described by an essentially electrostatic model in which the
primary interaction involves the electrophilic end $H^{\delta+}$ of HX and a
nucleophilic n-pair on B. But necessarily, if one part of B is nucleophilic

then another part will be relatively electrophilic and similarly for HX, *i.e.* each molecule is amphiphilic. Thus, in general there will be more than one interaction of the nucleophile-electrophile type involving B and HX. Can we predict which will be the primary (*i.e.* lowest energy) interaction? To consider this question, we shall use the simple example of the two isomers $H_2O\cdots HX$ and $HX\cdots HOH$.

It has been possible to assign a gas-phase nucleophilicity N and an electrophilicity E to the appropriate regions of each of H_2O and HX by partitioning the intermolecular stretching force constant k_σ of dimers $B\cdots HX$ between the components. After a large number of $B\cdots HX$ had been investigated, a systematic relationship among the k_σ was evident.[29] In fact, by assigning a number N to the acceptor region of each component B and similarly a number E to the donor region of each HX, all k_σ could be reproduced with acceptable accuracy through

$$k_\sigma = cNE \tag{1}$$

A selection of N and E values appropriate to the choice of $c = 0.25$ Nm^{-1} is given in Table 2. The molecules chosen have appeared in dimers both as acceptors and donors and hence each has both an N and and E value.[29]

Table 2.   Electrophilicities E and nucleophilicities N of some amphiphilic molecules

molecule	E	N
HF	10.0	4.8
HCl	5.0	3.1
H_2O	5.0	10.0
HCN	4.25	7.3
HCCH	2.4	5.1

Since k_σ is a measure of the strength of the interaction, it seems not unreasonable that it measures in some way the propensity of the acceptor molecule to seek an electrophile and of the donor molecule to seek a nucleophile. Equation (1) is understandable on this basis. In an extension of the argument, the relative magnitudes of the products $N_B E_{HX}$ and $N_{HX} E_B$ will give information about whether, for example, $H_2O\cdots HX$ or $HX\cdots HOH$ is the more stable. Presumably, the isomer with the larger product NE will be the lower in energy. Table 3 compares

Table 3.   Products NE for $H_2O\cdots HX$ and $HX\cdots HOH$

HX	HF	HCl	HCN	HCCH	HOH
$N_{H_2O} E_{HX}$	100.0	50.0	42.4	24.0	50.0
$N_{HX} E_{H_2O}$	24.0	15.6	36.4	25.6	50.0

these products for the two forms $H_2O\cdots HX$ and $HX\cdots HOH$ for a range of groups X. We note that according to this criterion while the isomer

$H_2O \cdots HF$ is strongly favoured with respect to $HF \cdots HOH$, the order of stability of the isomers has changed when HX is ethyne. The two isomers $H_2O \cdots HCCH$ and $HCCH \cdots HOH$ are in fact predicted to have comparable stability. Although only $H_2O \cdots HCCH$ has so far been observed[30] in low-temperature supersonic expansions of $H_2O/HCCH$ in Ar, it seems possible that the other isomer might also be present in detectable population. When the two products $N_B E_{HX}$ and $N_{HX} E_B$ are both small it is possible that the molecules will choose a relative orientation that is a compromise between the two weakly bound extremes and that allows additional stability through two interactions. $H_2CO \cdots HCCH$ appears to be a case of this type[28] where the amphiphilic nature of both components is used to give a more stable interaction in a single isomer (see Section 6). Presumably, the extra stability achieved through the additional interaction compensates for the small energy required to bend the very weak $H_2CO \cdots HCCH$ bond.

8   HYDROGEN BONDS B$\cdots$HX WHEN B CARRIES NO n-PAIRS

We have seen that the simple rule (namely, in the equilibrium conformation of B$\cdots$HX the HX molecule lies along the axis of an n-pair carried by B) is apparently electrostatic in origin in the sense that the electrophilic region $H^{\delta+}$ of HX seeks the most nucleophilic region of B. But what determines the angular geometry when B has no n-pairs?

Complexes B$\cdots$HX where B carries π-pairs or pseudo π-pairs

The simplest examples of molecules B that have π- or pseudo π-pairs but no n-pairs are ethyne, ethene and cyclopropane. For complexes B$\cdots$HX involving these molecules, we might expect, by analogy with the n-pair rule, to find a rule which requires the $H^{\delta+}$ of HX to seek the symmetry axis of a π or pseudo-π orbital since these are now the directions of greatest nucleophilicity. The experimentally determined angular geometries[31,32] of two dimers B$\cdots$HCl involving a π-type hydrogen bond are shown in Figure 12. It is clear that both geometries are as expected from the modified rule. Figure 12 also shows the Coulson-Moffitt model[33] of the pseudo-π bond in cyclopropane formed by sp^3 - sp^3 overlap on adjacent C atoms and the observed angular geometry[34] of cyclopropane$\cdots$HCl. The HCl molecule lies along the extension of a median of the equilateral triangle, as expected from the model.

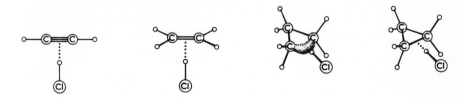

Figure 12

The order of the nucleophilicity N of the π- and pseudo-π pairs is of interest. The values, determined as discussed in Section 7 from k_σ, are[35]:

π-pairs:	cyclopropane	ethyne	ethene
N	6.4	5.1	4.7

The order for n-pairs in a range of simple molecules B is[29]

n-pairs:	NH_3	H_2O	HCN	PH_3	CO	N_2
N	11.5	10.0	7.3	4.4	3.4	2.2

We note that the pseudo-π-pair of cyclopropane is a better nucleophile than π-pairs in ethyne or ethene and indeed has an N value not far from that of the n-pair in HCN. The order N(ethyne)> N(ethene) has been tested in an examination of the angular geometry of 1-buten-3-yne···HCl. The result (Figure 13)[36] shows that, in the lowest energy isomer, HCl forms a hydrogen bond to a π-pair of the C≡C bond rather than the C=C bond, with the angle φ~34°.

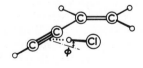

Figure 13

Do σ-bonding electron pairs act as proton acceptors?

If the pseudo-π electron pair in cyclopropane can act as a proton
acceptor in forming a hydrogen bond, can the σ-bonding pair in ethane
do likewise? We recently examined the rotational spectrum of a dimer of
ethane with hydrogen cyanide.[37] The symmetric-rotor type spectrum
and the frequency shifts that accompany H/D substitution in the HCN
subunit establish unambiguously that the geometry of the dimer is of the
C_{3v} type shown in Figure 14. Thus the HCN molecule forms a (weak)
hydrogen bond to the centre of the face of a methyl group. Presumably,
this region is more nucleophilic than the σ-bonding region between the
carbon atoms. Similar hydrogen bonds to methyl group faces have been
found in methane···HX complexes.[38-40] Such interactions, although
weak ($N = 2.3$ for ethane), might be a factor when considering
interactions between pairs of large, multisite molecules.

Figure 14

9   PROTON TRANSFER IN HYDROGEN-BONDED DIMERS

In polar sovents, proton transfer is an important and familiar
phenomenon. A question of interest concerns the possibility of proton
transfer (or partial transfer) when not mediated by a polar solvent.
Under what conditons can a proton be transferred from HX to B when
the two molecules interact in the absence of polar solvent or any solvent?
An answer is available from recent investigations of the ammonium
and methylammonium halides through their rotational spectra.[41-47]

It is notoriously difficult to place a hydrogen atom through
rotational spectroscopy using H/D substitution, especially if the proton in
question lies close to the centre of mass of a dimer B···HX. Thus another
method of distinguishing between the extreme possibilities, *e.g.*
$(CH_3)_3N$···HX and $(CH_3)_3NH^+$···X^-, must be found. Two methods have in
fact been used.

The first employs the H,F nuclear spin-nuclear spin coupling
constants of dimers B···HF, as determined from analyses of hyperfine
structure in rotational transitions. This coupling is the magnetic
through-space effect that is averaged to zero in solution-phase NMR
spectroscopy and its strength varies as $<r^{-3}>$, where r is the H-F

internuclear distance. A comparison of the coupling constants of B···HF and free HF allows an estimate of the lengthening δr of the HF bond when incorporated into the dimer. The following values of δr have been determined in this way for the indicated series of B···HF dimers.[12,43]

B	N_2	CO	HCN	CH_3CN	H_2O	$(CH_3)_3N$
$\delta r/\text{Å}$	0.001	0.007	0.014	0.016	0.015	0.041
k_σ/Nm^{-1}	5.5	8.5	18.2	20.1	24.9	39

We conclude that the extent of proton transfer is negligibly small, even for $(CH_3)_3N$···HF in which the hydrogen bond is very strong but for which $\delta r/r$ is only 0.05. This result is not unexpected in view of the known extreme resistance of the HF bond to extension. When can we expect proton transfer in the gas phase? Presumably, it is most likely when B has a large proton affinity while the energy required for the dissociation HX = H^+ + X^- is as small as possible.

The gas-phase proton affinity of NH_3 is large and increases progressively as ammonia is methylated. The energy required to dissociate HX into H^+ and X^- decreases along the series X = F > Cl > Br > I. These facts suggest examination of series H_3N···HX in which NH_3 is progressively methylated. The quantities from the rotational spectra of ammonium and methylammonium halides (X=Cl, Br, I) that are sensitive to the position of the hydrogen bond proton are the halogen nuclear quadrupole coupling constants and the stretching force constants k_σ. For example, in the covalent molecule $H^{35}Cl$ and the ion pair Na^+ $^{35}Cl^-$, the Cl- coupling constant (which leads to the electric field gradient at the Cl nucleus along the molecular symmetry axis) varies[45] from -67.62 MHz to -5.643 MHz. On the other hand $k_\sigma \sim$ 10-20 Nm^{-1} for a typical hydrogen-bonded dimer but $\sim$ 100 Nm^{-1} for an ion pair like Na^+···Cl^-. Examination[41,44,45] of the variation of the Cl-nuclear quadrupole coupling constant and k_σ along the series H_3N···HCl, CH_3NH_2···HCl, $(CH_3)_3N$···HCl reveals a systematic decrease in the magnitude of the former and a concomitant increase of the latter. This indicates that, although H_3N···HCl should be described as a simple hydrogen-bonded complex, partial proton transfer should be invoked for $(CH_3)_3NH^+$···Cl^- even in the gas phase. A similar investigation[46] of the analogue $(CH_3)_3NH^+$···Br^- leads to the conclusion that the extent of proton transfer is greater and that this species is largely an ion pair. The extent of proton transfer therefore increases along the series $(CH_3)_3N$···HF < $(CH_3)_3N$···HCl < $(CH_3)_3N$···HBr. Conversely, for the series H_3N···HCl, H_3N···HBr and H_3N···HI no significant transfer is apparent.[41,42,47]

REFERENCES

1. A.C. Legon, D.J. Millen and S.C. Rogers, Proc. R. Soc. London. Ser. A, 1980, 370, 213.
2. A.C. Legon, Annu. Rev. Phys. Chem., 1983, 34, 275.
3. A.C. Legon and D.J. Millen, Chem. Rev. 1986, 86, 635.
4. J.W. Bevan, Z. Kisiel, A.C. Legon, D.J. Millen and S.C. Rogers, Proc. R. Soc. London. Ser. A, 1980, 372, 441.
5. Z. Kisiel, A.C. Legon and D.J. Millen, Proc. R. Soc. London. Ser. A, 1982, 381, 419.
6. Z. Kisiel, A.C. Legon and D.J. Millen, J. Chem. Phys., 1983, 78, 2910.
7. A.C. Legon and L.C. Willoughby, Chem. Phys. Lett., 1982, 92, 333.
8. Z. Kisiel, A.C. Legon and D.J. Millen, J. Mol. Struct., 1984, 112, 1.
9. Z. Kisiel, A.C. Legon and D.J. Millen, J. Mol. Struct., 1985, 131, 201.
10. A.C. Legon, D.J. Millen and H.M. North, Chem. Phys. Lett., 1987, 135, 303.
11. G. Cazzoli, P.G. Favero, D.G. Lister, A.C. Legon, D.J. Millen and Z. Kisiel, Chem. Phys. Lett. 1985, 117, 543.
12. A.C. Legon and D.J. Millen, Proc. R. Soc. London. Ser. A, 1986, 404, 89.
13. A.C. Legon and D.J. Millen, Discuss. Faraday Soc., 1982, 73, 71.
14. A.C. Legon and D.J. Millen, Chem. Soc. Rev., 1987, 16, 467.
15. R. Viswanathan and T.R. Dyke, J. Chem. Phys., 1982, 77, 1166.
16. F.A. Baiocchi and W. Klemperer, J. Chem. Phys., 1983, 78, 3509.
17. G.J.B. Hurst, P.W. Fowler, A.J. Stone and A.D. Buckingham, Int. J. Quantum Chem., 1986, 29, 1223.
18. A.J. Fillery-Travis and A.C. Legon, J. Chem. Phys., 1986, 85, 3180.
19. A.D. Buckingham and P.W. Fowler, Can. J. Chem., 1985, 63, 2018.
20. R.A. Collins, A.C. Legon and D.J. Millen, J. Mol. Struct., 1987, 162, 31.
21. A.S. Georgiou, A.C. Legon and D.J. Millen, J. Mol. Struct., 1980, 69, 69.
22. A.S. Georgiou, A.C. Legon and D.J. Millen, Proc. R. Soc. London. Ser. A, 1980, 373, 511; A.C. Legon, D.J. Millen and A.L. Wallwork, Chem. Phys. Lett., 1991, 178, 279.
23. A.C. Legon and C.A. Rego, Angew. Chem. Int. Ed. Engl., 1990, 29, 72.
24. A.C. Legon and A.L. Wallwork, J. Chem. Soc. Faraday Trans., 1990, 86, 3975.
25. A.C. Legon, A.L. Wallwork and H.E. Warner, J. Chem. Soc. Faraday Trans., 1991, 87, 3327.
26. A.C. Legon and D.J. Millen, Chem. Soc. Rev., 1992, 21, 71.
27. E.J. Goodwin and A.C. Legon, J. Chem. Phys., 1987, 87, 2426.
28. N.W. Howard and A.C. Legon, J. Chem. Phys., 1988, 88, 6793.
29. A.C. Legon and D.J. Millen, J. Am. Chem. Soc., 1987, 109, 356. The value of E_{H_2O} has been derived from $k_\sigma = 11.7$ Nm^{-1} for H$_2$O$\cdots$HOH, as reported by T.R. Dyke, K.M. Mack and J.S. Muenter, J. Chem. Phys., 1977, 66, 498.
30. G.T. Fraser, K.R. Leopold and W. Klemperer, J. Chem. Phys., 1983, 80, 1423.
31. A.C. Legon, P.D. Aldrich and W.H. Flygare, J. Chem. Phys., 1981, 75, 625.
32. P.D. Aldrich, A.C. Legon and W.H. Flygare, J. Chem. Phys., 1981, 75, 2126.
33. C.A. Coulson and W.E. Moffitt, Phil. Mag., 1949, 40, 1.
34. A.C. Legon, P.D. Aldrich and W.H. Flygare, J. Am. Chem. Soc., 1982, 104, 1486.
35. A.C. Legon and D.J. Millen, J. Chem. Soc., Chem. Commun., 1987, 986.
36. Z. Kisiel, P.W. Fowler, A.C. Legon, D. Devanne and P. Dixneuf, J. Chem. Phys., 1990, 93, 6249.
37. A.C. Legon, A.L. Wallwork and H.E. Warner, Chem. Phys. Lett., 1992, 191, 97.
38. A.C. Legon, B.P. Roberts and A.L. Wallwork, Chem. Phys. Lett., 1990, 173, 107.

39.   A.C. Legon and A.L. Wallwork, J. Chem. Soc. Faraday Trans., 1992, 88, 1.
40.   M.J. Atkins, A.C. Legon and A.L. Wallwork, Chem. Phys. Lett., in press.
41.   N.W. Howard and A.C. Legon, J. Chem. Phys., 1988, 88, 4694.
42.   N.W. Howard and A.C. Legon, J. Chem. Phys., 1987, 86, 6722.
43.   A.C. Legon and C.A. Rego, Chem. Phys. Lett., 1989, 154, 468 and 1989, 157, 243.
44.   A.C. Legon and C.A. Rego, J. Chem. Phys., 1989, 90, 6867.
45.   A.C. Legon and C.A. Rego, Chem. Phys. Lett., 1989, 162, 369.
46.   A.C. Legon, A.L. Wallwork and C.A. Rego, J. Chem. Phys., 1990, 92, 6397.
47.   A.C. Legon and D. Stephenson, J. Chem. Soc. Faraday Trans., 1992, 88, 761.

An Approach to Molecular Recognition Based on Partitioning of Free Energy Contributions

Dudley H. Williams and Mark S. Searle

UNIVERSITY CHEMICAL LABORATORY, LENSFIELD RD., CAMBRIDGE CB2 1EW, UK

1 INTRODUCTION - A PARTITION OF THE FREE ENERGY OF BINDING; EQUATIONS FOR THE ESTIMATION OF BINDING CONSTANTS

Following the pioneering work of Jencks[1], and Page and Jencks[2], in two recent publications[3,4] we have factorised the free energy of binding for a molecular association in aqueous solution into four terms. A similar factorisation has previously been used by Andrews et al.,[5] and the relevance and physical basis of the factors involved have been summarised by Fersht.[6] The consideration of only four terms is justified only if the ligand and receptor show good van der Waals complementarity, and if the conformations of the bound components correspond closely to conformational energy minima in the separated states.[3,4] These terms are considered as follows: (i) the low probability of "catching" the ligand on the receptor in the absence of intermolecular forces; (ii) the adverse free energy change (largely entropic) associated with the restriction of any internal rotations of either component upon complex formation; (iii) the promotion of binding if hydrocarbon is removed from exposure to water upon complex formation; and (iv) the promotion of binding due the favourable interactions between polar functional groups in the complex. These four parameters are now enumerated and elaborated. Problems associated with the designation of particular values to the parameters are then discussed.

(i) **Bimolecular Association:** Any bimolecular binding process is entropically unfavourable due to the formation of a single molecule of complex, which occurs with loss of translational and rotational entropy. The unfavourable free energy of the association (ΔG_{t+r}, kJ mol^{-1}, equal to $-T\Delta S_{t+r}$ at 298K) as a function of the molecular weight of a ligand binding to a larger receptor with complete loss of the translational and rotational entropy of the smaller component can be estimated,[3] and is given in Figure 1.[7] As with all other free energy changes given in this account, it can be converted to an effect on $\log_{10}K$ by division by 5.7 (for room temperature binding); this scale is given on the right hand side of the Figure. We find that within 4 kJ mol^{-1}, the same values apply for any molecular shape (rod, disc, or sphere) of molecular weight m binding to any receptor of molecular mass 1200 or greater.[3] Thus, for example, ΔG_{t+r} is adverse to binding by a factor of ca.10^{-9} M^{-1} for a ligand of molecular weight 200 (where the "molecular weight" includes bound solvent molecules, which can be regarded as translating and rotating with the ligand). The figure of 10^{-9} M^{-1} is the entropic price to be paid when there is no residual overall relative translation and rotation of the associating components **A** and **B** in the complex **A.B**. We note later that this situation is unlikely to be achieved in practice; rather, some (variable) fraction of 10^{-9} M^{-1} is the cost to be paid.

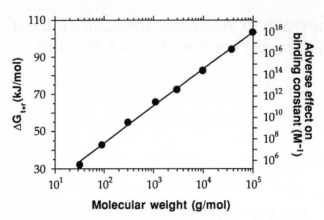

Figure 1. An estimate of the adverse effect on binding constant in the hypothetical case where all the entropy of translation and overall rotation of one (the smaller) component is lost upon association.

(ii) **Restriction of Internal Rotations:** Following Page and Jencks,[2] we note that binding is adversely affected by approximately 5 to 6 kJ mol^{-1} (ΔG_r) for each rotation removed upon association. Thus, if four relatively free rotations are lost in an association, the binding is hypothesised to be adversely affected by a factor of ca 10^4. These values correspond to severe restriction, or complete loss of a rotation, and we argue later that smaller values may be appropriate for associations involving non-covalent bonds.

(iii) **The Hydrophobic Effect**: For every square Angstrom ($\mathring{A}^2$) of hydrocarbon removed from exposure to water by the binding process, we assume the binding energy to be increased by 0.2 kJ mol^{-1}.[4,8] This value is based on thermodynamic measurements of the solubility of simple hydrocarbons in water, which indicate that this "hydrophobic effect" is essentially entropy driven at room temperature. Thus, if the area of hydrocarbon buried is x$\mathring{A}^2$, then the free energy change (ΔG_h) due to the hydrophobic effect is taken as 0.2x kJ mol^{-1}. An alternative approach is to use the data of Fersht[6] on many different enzymes. If his values for the binding energies of CH_3, CH_2CH_3, and $CH(CH_3)$ groups are reduced by 8 kJ mol^{-1} per methylene group (as this is the additional binding energy when a methylene group fills a "hole" in an enzyme structure,[6] and is not relevant to open binding sites), then a hydrophobic effect of 2.5 kJ mol^{-1} per CH group is obtained. In using such values, a methyl group would be regarded as 3 CH groups. Note that this value accounts also for the removal from water of the hydrocarbon against which the CH groups bind.[6]

(iv) **Polar Interactions of Functional Groups:** The bringing together of the two binding entities with the appropriate internal geometry is accounted for in factors (i) and (ii). However, if the free energy of binding which results from the interaction of any pair of functional groups (ΔG_p) is to be the same if the process occurs either intramolecularly or bimolecularly (which is obviously desirable if ΔG_p values are to be of general utility), then ΔG_{t+r} values cannot be taken from Figure 1. We elaborate on this point subsequently. When polar interactions occur with optimum geometry for binding, they are known as intrinsic binding energies.[1]

In summary, the free energy (ΔG, kJ mol^{-1}) of a bimolecular association following the above specifications might be approximated by:

$$\Delta G = \Delta G_{t+r} + \Delta G_r + \Delta G_h + \Sigma \Delta G_p \qquad \text{Equation 1}$$

where $\Sigma \Delta G_p$ represents the free energies of binding for each set of interacting functional groups, summed over all such sets of interactions.

For the more general case where ΔG_{conf} represents the total conformational strain energy produced upon binding, and ΔG_{vdW} represents the change in van der Waals energy between free and bound states (due, for example, to the existence of van der Waals repulsions or cavities in the complex), then equation 2 results[3]:

$$\Delta G = \Delta G_{t+r} + \Delta G_r + \Delta G_h + \Sigma \Delta G_p + \Delta G_{conf} + \Delta G_{vdW} \quad \text{Equation 2}$$

2 FUSION AND SUBLIMATION: MODELS FOR COMPLEX DISSOCIATION

Crystalline substances provide a model of a biological complex in which motions due to overall translation and rotation, as well as freedom of internal bond rotations, are restricted by comparison with those in a gas, pure liquid, or a solution. Since organic crystals are held together by relatively weak intermolecular (rather than covalent) forces, their formation from melts can perhaps give useful guides to corresponding changes in $T\Delta S_{t+r}$ and $T\Delta S_r$ (which correspond to ΔG_{t+r} and ΔG_r, respectively) for the formation of weakly bound complexes. The values of ΔG_{t+r} and ΔG_r presented in the previous section are largely based on changes involving the formation of covalent bonds. The important question is "how different are these values for the formation of weakly bound complexes than for those involving covalent bond formation?"

Consider first the process of fusion, for which the entropy of fusion (ΔS_f) is related to the enthalpy of fusion (ΔH_f) by the melting temperature ($T_m, {}^\circ K$):

$$\Delta S_f = \Delta H_f / T_m$$

ΔS_f reflects an increase in entropy because the amplitudes of overall molecular translations and rotations (ΔS_{t+r}) and any possible internal bond rotations (ΔS_r) are increased upon melting, but also a decrease in entropy because favourable soft vibrations are reduced (ΔS_v) in the transition from solid to melt. Since entropies of fusion are always positive, it is clear that the former effects are always greater than the latter, identifying an entropic cost in forming a crystal (cf. forming a complex). We can describe ΔS_f as a balance of these terms:

$$\Delta S_f = \Delta S_v + \Delta S_{t+r} + n\Delta S_r \qquad \text{Equation 3}$$

The observation that enthalpies of vapourisation are frequently 2 to 12 times larger than enthalpies of fusion suggests that molecules in the melt are still partially ordered by intermolecular forces between neighbours; for this reason, we expect the terms on the right hand side of equation 1 to have smaller values than for complex dissociations in solution.

First, we consider the relationship between the entropy of fusion and the number of rotors for the linear hydrocarbons, as plotted in Figure 2.[9] For a C_n hydrocarbon chain, n-3 rotors are restricted in the crystallisation process. n-Butane is considered to have only a single rotor; the methyl groups have small barriers to rotation in the solid and are regarded as rotating in this state.

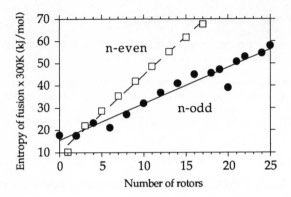

Figure 2: Entropies of fusion of odd- and even-membered linear saturated hydrocarbons as a function of the number of internal rotors.

The plots of $T\Delta S_f$ (at 300K) versus n (up to 25) for odd and even series carbon chains reveal good linear correlations (R>0.97) with the slopes providing an estimate of the cost of restricting a backbone rotation in the two types of hydrocarbon chain as 1.6 and 3.4 kJ mol^{-1}. The distinction between odd and even series up to C_{20} is made on the basis of crystal type and the occurrence of solid-solid phase transitions that are observed only for the odd series. Thus, just below the melting temperature the odd series hydrocarbons are less well-ordered and this is reflected in a smaller favourable entropy change per rotor on melting. Extending the analysis to homologous series of alkyl carboxylic acids and methyl ketones:

good linear correlations are again found (R>0.95), giving the entropic cost of restricting a rotor ($T\Delta S$ at 300 K) as 2.3 and 3.6 kJ mol^{-1}, respectively.[9] Even the largest of these numbers is less than the 5.5 kJ mol^{-1} per rotor derived by Page and Jencks for "freezing" a rotor, reflecting the fact that molecules held by weak interactions probably have greater residual motions than for molecules involved in the covalent transformations reported by Page and Jencks.[2] The observation of a linear relationship between $T\Delta S_f$ and the number of rotors suggests that for a given series $[T\Delta S_{t+r} + T\Delta S_v]$ is approximately constant.

We have considered the entropies of fusion of a large number of organic compounds (largely free of internal rotors) and find a striking relationship between the entropy and enthalpy of fusion for molecules of similar mass, as plotted in Figure 3.[9] Large enthalpies generally correlate with more polar molecules, but those with large negative enthalpies of crystallisation pay the price by having a large unfavourable entropy of crystallisation. The enthalpies and entropies of fusion of molecules with relatively high symmetry, such as cyclopentane, reflect highly disordered crystals with very small enthalpies and entropies of crystallisation. The plot in Figure 3 suggests that the largest possible adverse entropy change, corresponding to complete immobilisation in the crystal [to which the ΔG_{t+r} (= $-T\Delta S_{t+r}$) in Figure 1 corresponds], is unlikely to be achieved even in a tight complex (i.e. very large enthalpy of crystallisation), and suggests the importance of the favourable entropy of residual motions in molecular recognition complexes.

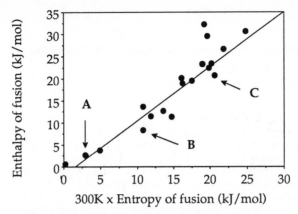

Figure 3: Correlation between entropy and enthalpy of fusion for a range of organic compounds of varying polarity: A(cyclohexane); B(pyridine); C(2-nitro-5-methylphenol).

The same compensatory relationship between enthalpy and entropy is clearly evident in the data on sublimation. In general, entropies of sublimation amount to only 40-70% of the experimental or calculated gas phase entropies, the rest being attributed to residual motions in the crystal. Thus, if crystals represent adequate models for the formation of complexes which are important in biology, then the adverse entropy of bimolecular associations may be much less than that leading to an effect of ca 10^{-9} to 10^{-10} M^{-1} on association rates where association leads to covalent bond formation.

3 ESTIMATING ENTROPY CHANGES FOR MOLECULAR ASSOCIATIONS IN SOLUTION

The possibility of partial ordering of molecules in the melt suggests that entropies of fusion are probably less than those for the dissociation of complexes in solution; the entropies of gases are of course significantly larger than those of liquids, making entropies of sublimation an overestimate. However, the sublimation model has the advantage that molecules are likely to be fully dissociated in the gas, as they might be in solution. The observation that the entropy of condensation for a large number of liquids at their boiling points is found to be approximately constant at -87 J K^{-1} mol^{-1} (Trouton's rule) enables us to correct the entropy of sublimation data to a value more representative of entropy changes for complex dissociation in solution. When two additional small corrections are included to account for cooling to 300 K and for dilution to a 1 M solution, we obtain a correction of -29 kJ mol^{-1} (TΔS at 300 K) as a rough guide to the reduction which should be applied to gas phase entropies to give solution entropies prior to association. Using this approximation for the cases of large exothermicities of crystallisation, the estimated entropic cost of binding is relatively large (very roughly 45 kJ mol^{-1}), and comparable to the accepted range of values for the reduction in entropy proposed by Page and Jencks (50-70 kJ mol^{-1}, involving the use of models where reactions occur with covalent bond formation[2]). However, we conclude that associations with small exothermicities are associated with smaller entropy changes for complex formation in solution. In some cases, the benefit of translational and rotational entropy which remains in the complex, plus the entropic advantage of new low frequency vibrations, may almost balance the formal loss of $T\Delta S_{t+r}$ (close in magnitude to the ΔG_{t+r} values which are given in Figure 1) associated with the formation of a rigid complex. This can be illustrated for the dimerisation of 2-pyridone in chloroform[10]:

The formation of a rigid dimer opposes binding (by ca 10^{-10} M^{-1}) in binding constant due to loss of translational and rotational entropy. The experimental adverse entropy (in terms of $T\Delta S$) of only $10^{-2.3}$ M^{-1} indicates that the advantage of residual motions and "sloppy" hydrogen bonds (including new NH---OC vibrations) promotes binding by ca $10^{7.7}$. Together with the favourable enthalpy term for the formation of two hydrogen bonds of $10^{4.4}$, a large favourable overall free energy change for dimerization of ca 10^2 M^{-1} is found. Note that this analysis is in accord with the relatively small entropic cost of the weakly exothermic crystallisation of hydrocarbons (only -10 to -20 kJ mol^{-1} in the absence of freeing internal rotors, from the intercept on the entropy axis in Figure 2); and with the small adverse entropy of crystallisation of organic compounds where the exothermicity is low (Figure 3).

The implications of these conclusions for the application of equation 1 are now considered.

4 UNCERTAINTIES IN THE APPLICATION OF EQUATION 1

ΔG_{t+r}: The values which can be read off from Figure 1 are not experimental and solution entropies of complex molecules, upon which they depend, are uncertain. However, the values do accord well with the adverse entropy changes (on which they largely depend) for reactions in which two small molecules react to one **covalently** bound entity.[1]

The Entropic Cost of Bimolecular Associations through Weak Interactions: A much more difficult problem than that outlined in the previous paragraph lies in searching for the most acceptable method to do the "accounting" for the free energy change, ΔG_{t+r}, appropriate for weak associations. The section dealing with using the entropies of fusion and sublimation of crystals as models for entropy changes in weak associations highlights the problem. Not only may the value of ΔG_{t+r} read off from Figure 1 be much larger than the loss in translational and rotational free energy for a weak association, but the degree by which it is too large seems likely to be dependent on the exothermicity of the association. Weakly exothermic associations are expected to have relatively small adverse entropies, and strongly exothermic associations to have much larger adverse entropies.

As noted previously, the values in Figure 1 correspond to making a complex in which all the relative overall translation and rotation of the associating entities has been lost. Thus, if the values in Figure 1 are used directly in the application of equation 1 to an association $A + B \rightarrow A.B$, then entropically advantageous relative translation and rotation of A and B which remain in $A.B$ are credited to the functional group interactions in $\Sigma\Delta G_p$. Using this approach, and the data of Susi et al.[11] for valerolactam dimerisation in aqueous solution, ΔG_p for each amide-amide hydrogen bond in the dimer (1) is -27 kJ mol^{-1}.[12] In broad terms, this is because the reported association constant for dimerisation is ca 10^{-1} M^{-1}, and the loss of the translational and rotational entropy of one component of the dimer is adverse to binding by ca. 10^{-9} M^{-1} (cf. Figure 1). Thus, the amide-amide hydrogen bonds are concluded to promote dimerisation by ca. 10^8 M^{-1}, or ca. 10^4 M^{-1} per hydrogen bond. The entropically favourable low frequency vibrations which permit relative translation of the two parts of the dimer, and other entropically favourable motions in the dimer have been credited to the ΔG_p values of the hydrogen bonds.

An alternative approach is not to use ΔG_{t+r} values taken from Figure 1, but smaller values which therefore treat motions remaining in the complex as constituting translational and rotational entropy that was not lost. A good case can be made that the loss in translational and rotational entropy in the formation of the dimer 1 is in reasonable accord with the expectations from the fusion and sublimation models. Murphy and Gill[13] have shown that the entropy change (in terms of TΔS at 298K) upon dissolving crystals of diketopiperazine in aqueous solution is +11.5 kJ mol^{-1}. If allowance is made for the fact that the two methylene groups of this molecule will order water (through the hydrophobic effect) in aqueous solution, causing an adverse TΔS on dissolution of 5-10 kJ mol^{-1}, then freeing the molecule from the crystal is seen to be favourable entropically (TΔS) by 16.5-21.5 kJ mol^{-1}. In other words, in passing from free diketopiperazine in aqueous solution to its crystal lattice (a model for a bimolecular association involving similar weak interactions), the molecule loses only about 33-43% of its translational and rotational free energy (i.e., about these percentages of the value read off from Figure 1). Since diketopiperazine is held in its crystal structure by 4 amide-amide hydrogen bonds, there can be little doubt that the proposed solution dimer 1 would lose an even smaller fraction of ΔG_{t+r}. Plausibly, the adverse change in ΔG_{t+r} (-TΔS_{t+r}) in the formation of the dimer 2 would lie in the range 11-17 kJ mol^{-1}, opposing dimer formation by a factor of only 10^{-2} to 10^{-3} M^{-1}. Thus, if ΔG_p= -(5 to 6) kJ mol^{-1} (promotion of dimerisation by a factor of 10^1 M^{-1}) for the amide-amide hydrogen bond, two of these would be adequate to overcome the adverse entropy of dimerisation to give an observed dimerisation constant in the region of 10^{-1} M^{-1}. In summary, large negative ΔG_p values are obtained for the hydrogen bond if it is credited with the favourable entropy of residual motions, and small negative ΔG_p values if it is not credited with this favourable entropy.

In relation to the above discussion, it is noteworthy that the evidence for the formation of the dimer 1 in aqueous solution, although long accepted, is not particularly strong. Against this point may be set the similar dimerisation constants (ca. 10^{-1} M^{-1}) proposed for *assumed* analogous dimers of urea[14] and diketopiperazine[15] in aqueous solution; and the fact that dimerisation constants in this region are consistent with the ΔG_p values (-5 to -6 kJ mol^{-1}) assumed for amide-amide hydrogen bonds on the basis of protein engineering experiments (which do not credit this hydrogen bond with the advantageous entropy of the large residual motions which its formation allows - see below). Evidence for the formation of hydrogen-bonded dimers of γ-butyrolactam, δ-valerolactam, and ε-caprolactam in carbon tetrachloride is much stronger. The respective values of K, ΔH, and TΔS at 298K are: γ-butyrolactam 460 M^{-1}, -29 kJ mol^{-1}, and -14 kJ mol^{-1}; δ-valerolactam 432 M^{-1}, -43 kJ mol^{-1}, and -28 kJ mol^{-1}; ε-caprolactam 168 M^{-1}, -23 kJ mol^{-1}, and -10 kJ mol^{-1}.[16] These data give support to the model which uses crystal formation as a guide to the entropic costs of complex formation. From the dimerisation constants, it can be seen that the ΔG values for dimer formation are all similar (-15, -15, and -13 kJ mol^{-1}, respectively). Thus, where dimer formation is most exothermic (δ-valerolactam), the *extent* of dimer formation is similar to the other cases. This is because the most exothermic dimerisation pays the largest cost in entropy. Covering all three lactams, the entropic costs of dimerisation, in terms of the the adverse cost of TΔS at 298K on the dimerisation constant, lie in the range $10^{-1.7}$ to $10^{-4.7}$ M^{-1}. The smallest entropic cost is for the least exothermic association, and the largest cost for the most exothermic. This observation, and the range of the entropic cost, support the guides given by the crystal model.

Studies Involving the Formation of the Amide-Amide Hydrogen Bond in a Ligand Extension; Interaction of ΔG_{t+r}, ΔG_r, and ΔG_h: An analysis of the increased binding constant of N-Ac-Gly-D-Ala (**2**) over N-Ac-D-Ala (**3**) to the antibiotics ristocetin A and vancomycin gave an average binding energy (ΔG_p) of -24 kJ mol^{-1} for the amide amide hydrogen bond found between the NH of glycine and an amide CO of the antibiotics (Figure 4).[3] This value was obtained by taking the mean increase in binding energy of **2** over **3** (11 kJ mol^{-1} at 25^{0}C), adding 10 kJ mol^{-1} to this value (for the free energy cost of restricting two rotors at 5 kJ mol^{-1}, and also allowing for the estimated extra entropic cost (3 kJ mol^{-1}) for catching a ligand of larger mass.

$$-\Delta G_p = -11 - 10 - 3 = -24 \text{ kJ mol}^{-1} \quad \text{Equation 4}$$

2   **3**

In the method of accounting used to derive this value, the value for rotor restriction (5 kJ mol^{-1}) may give too little credit for the favourable free energy of residual torsions in the restricted rotors of weak complexes. If the values for rotor restriction in the formation of crystals from melts of C_nH_{2n+2} hydrocarbons (where n is even) at T_m are used (3.5 kJ mol^{-1}, see preceding section)as a lower limit, but 5 kJ mol^{-1} retained as an upper limit [since rotor restrictions are more costly in free energy when at sp^2-sp^3 bonds (peptides) than at sp^3-sp^3 bonds (hydrocarbons),[2] then -10 is replaced by -(7 to 10) kJ mol^{-1}. Second, the factor of -3 kJ mol^{-1} (equation 1) is based upon the fact that the entropy of a ligand is dependent upon the logarithm of its mass.[3] In the method of accounting adopted, the amide-amide hydrogen bond is credited with the larger adverse entropy that has to be overcome in catching a ligand of larger mass. No such credit may accrue if it is considered that the ligands **2** and **3** retain some fraction of their free ΔG_{t+r} when associated (rather than losing all their free ΔG_{t+r}, and then crediting residual motions in the complex to ΔG_p values). The reason for this is that the ΔG_{t+r} value of the free ligand is dependent upon a term log(am), and the ΔG_{t+r} value of the bound ligand upon a term log(bm) (where a and b are constants, and m is the mass of the ligand). Therefore, the difference between the ΔG_{t+r} values for the free and bound states is dependent upon a term log(am) - log (bm) [i.e., log (a/b)], and becomes independent of mass. In summary, if the ΔG_p value for the amide-amide hydrogen bond is not given the credit for any internal motions remaining in the complex, then -24 kJ mol^{-1} (equation 4) becomes -(18 to 21) kJ mol^{-1}.

The above changes in ΔG_p simply reflect different methods for crediting the advantages of residual motion in the complex, in proportioning the free energy charges between ΔG_{t+r}, ΔG_r, and ΔG_p. However, in addition to these "accounting method" differences, we conclude that the original analysis underestimated the role of the hydrophobic effect in increasing the binding of 2 over 3. In the original analysis,[3] it was argued that since the N-acetyl methyl group of 2 points away from the antibiotic, this methyl group would not perturb the binding significantly. Therefore, no increase in the binding of 2 over 3 was attributed to the hydrophobic effect (ΔG_h, equation 1). We now believe this to be an error for two reasons. First, the aromatic ring of residue 7 of the antibiotics (Figure 4) can approach this methyl group in the complex. Second, CPK models and molecular graphics representations of the complex show that other C-H groups of the antibiotic interact with the polar parts of the CH_3CONH- extension added in passing from 2 to 3. Using MacroModel, and calculating water-accessible surface areas of hydrocarbon with a water radius of 1.4A and the option of a high density of points on a sphere,[17] the increase in non-polar surface area which is removed from water when 2 binds to ristocetin A (over 3 binding to ristocetin A) is 38 Å^2.[18] As $\Delta G_h = -0.2 \times 38 = -8$ kJ mol^{-1}, and applying equation 1, we then obtain $\Delta G_i = -(10 \text{ to } 13)$ kJ mol^{-1}.

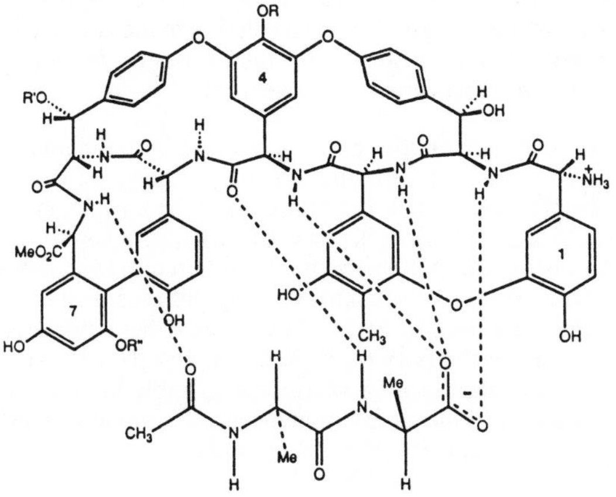

Figure 4. Exploded view of the complex formed between the antibiotic ristocetin A and the cell-wall analogue N-acetyl-D-Ala-D-Ala (with the aromatic ring 7 of the antibiotic indicated).

Using acetate ion as a fragment of 3 which binds to the antibiotic, ΔG_p for the amide-amide hydrogen bond which 3 forms to the antibiotics was determined as -16 kJ mol^{-1}.[3] This value credits ΔG_i with 6 kJ mol^{-1} for a greater ΔG_{t+r} of 3 over acetate anion and for residual motion reflected by using 5 vs. 3.5-5 kJ mol^{-1} per restricted rotor. The hydrophobic effect of the alanine methyl group of 3 was allowed for by an experimental value, but the hydrophobic interaction of the acetyl methyl group of 3 was taken as 2 kJ mol^{-1}. A more realistic value for this last interaction is 5 kJ mol^{-1} (since 2 CH groups of the methyl group interact with hydrocarbon of the antibiotic). Thus, when the credit for residual motion in the complex and for a larger hydrophobic effect is removed from -16 kJ mol^{-1}, we obtain $\Delta G_i = -(4 \text{ to } 7)$ kJ mol^{-1}. In summary, stepwise removal of the amide-amide hydrogen bonds which $\underline{2}$ and $\underline{3}$ make to vancomycin group antibiotics give binding energies of these hydrogen bonds in the range -(4 to 13) kJ mol^{-1}. Such are the combined uncertainties in the appropriate ΔG, ΔG_r, and ΔG_h values used in the application of equation 1, that the derived range cannot be regarded as being significantly different from the values -(2 to 8) kJ mol^{-1} for several types of neutral-neutral hydrogen bonds obtained from protein engineering experiments.[19,20] It is important to note that the ΔG_p values obtained for the binding of $\underline{2}$ and $\underline{3}$ to the antibiotics should probably bear comparison with the generalisation of -2 to -8 kJ mol^{-1}, since in protein engineering experiments local motions in the protein are probably little changed before and after a hydrogen bond deletion resulting from a mutation. Thus, the values give little or no entropic credit to the hydrogen bond for any motions it allows.

5 CONCLUSION

Large ΔG_p values for the amide-amide hydrogen bonds reportedly involved in dimerisations in aqueous solution are obtained if such hydrogen bonds are given credit for the residual motions which they allow. If the entropic advantage of these residual motions are estimated and removed, then the ΔG_p values obtained are not significantly different from the conventional view of these bonds. The same conclusion applies to ΔG_p values obtained by ligand extension (in binding to antibiotics) if credit for residual motions is removed, and allowance for a larger hydrophobic effect than originally envisioned is made. In reaching these conclusions, we also note that the evidence for the formation of hydrogen-bonded dimers in aqueous solution leaves much to be desired.

Efforts to estimate solution binding constants by the application of equations 1 and 2 will run into contradictions if ΔG_{t+r} values are taken from Figure 1, and the entropic advantages of residual motions thereby credited to ΔG_p values. These contradictions will arise because the *a priori* analysis of bimolecular associations of low exothermicity would credit much entropy of residual motions to ΔG_p values ($\underline{4}$). In contrast, if ligand extensions, or protein engineering experiments, are used to derive ΔG_p values, then entropically favourable motions are similar in both the ligand ($\underline{5a}$) and the extended ligand ($\underline{5b}$) if the extension is associated with a small or negligible increase in exothermicity of assocation. In these circumstances, ΔG_p values are not augmented by the favourable entropy of residual motion.

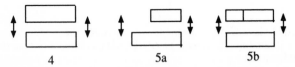

Therefore in attempts to obtain a potentially self-consistent set of ΔG_p values, the net change in ΔG_{t+r} should be employed. Much work remains to be done to determine this interesting parameter for various kinds of biologically important bimolecular associations. When crystals are used as models for complexes of biological interest, then the net loss in ΔG_{t+r} in associations is suggested to lie anywhere in the range 10 to 45 kJ mol^{-1} (opposing binding by a factor of $10^{-1.7}$ to 10^{-8} M^{-1}). Such values appear plausible from an analysis of the formation of lactam dimers in solution.

Acknowledgements: We thank SERC, the Upjohn Company (Kalamazoo), and Roussel Uclaf for financial suport.

REFERENCES

1. W.P.Jencks, Proc. Natl. Acad. Sci., 1981, *78*, 4046.
2. M.I. Page and W.P. Jencks, Proc. Natl. Acad. Sci., 1971, *68*, 1678.
3. D.H.Williams, J.P.L. Cox, A.J.Doig, M. Gardner, U. Gerhard, P.T. Kaye, A.R. Lal, I.A. Nicholls, C.J. Salter, and R.C. Mitchell, J. Amer. Chem. Soc., 1991, *113*, 7020.
4. D.H. Williams, Aldrichimica Acta, 1991, *24*, 71.
5. P.R. Andrews, D.J. Craik, and J.L. Martin, J. Med. Chem., 1984, *27*, 1648.
6. A.R. Fersht, "Enzyme Structure and Mechanism", W.H. Freeman, New York, Second Edition, 1985.
7. We have made three modifications in the estimation of ΔG_{t+r} relative to the original estimate (refs. 3 and 4). First, the Trouton's Rule correction refers to the entropy decrease of liquids at the boiling point relative to gases. To allow for the additional order existing at temperatures below the phase transition, we have added $T\Delta S$ to the Trouton's rule correction in water, where T is 298°C, and $\Delta S = C_p \ln(373/298)$, where C_p is the heat capacity of water. This correction is empirical, but seems better than ignoring an effect which is physically required in some form or another; it acts to reduce ΔG_{t+r} by 5 kJ mol^{-1}. Second, since the Trouton's rule correction is applicable to neat liquids, we had earlier (refs. 3 and 12) added $2.3RT\log_{10}1000/RMM$ to allow for the entropy increase at a molar concentration. Such a correction is appropriate for a concentration decrease in the gas phase of a single component, but not for a dilution in a liquid phase, where an entropy of mixing (assuming ideal behaviour) is appropriate. An entropy of mixing is small compared to the entropy of dilution, and for a ligand of effective molecular weight 200 Daltons, this change increases ΔG_{t+r} by 4 kJ mol^{-1}. Third, we originally followed the approach of Israelachvili that translational kinetic energy is not lost in an association to give a complex held together by weak intermolecular forces (Israelachvili, J.N. "Intermolecular and Surface Forces", Academic Press, London, 1985, p.21). In view of the case made later in this manuscript that considerable translational *and* rotational motions remain in complexes where the components are held together by non-covalent interactions, we now carry out the accounting by regarding kinetic energy of both translation and rotation (RT/2 per degree of freedom) to be retained in the complex; this modification increases ΔG_{t+r} by 4 kJ mol^{-1} (3RT/2). The overall effect of all these changes is to increase ΔG_{t+r} by only 3 kJ mol^{-1} relative to the original estimate[3,4] and, in view of the larger uncertainties in this parameter, these changes are cosmetic rather than of great consequence.
8. K.A. Sharp, A. Nicholls, R.F. Fine, and B. Honig, Science, 1991, *252*, 106.
9. M. Searle and D.H. Williams, unpublished work.
10. Y.Ducharme and J.D. Wuest, J. Org. Chem., 1988, *53*, 5787.
11. H. Susi, S.N. Timasheff, and J.S. Ard, J. Biol. Chem., 1964, *239*, 3051.
12. A.J. Doig and D.H. Williams, J. Amer. Chem. Soc., 1992, *114*, 338.
13. K.P. Murphy and S.J. Gill, Thermochimica Acta, 1990, *172*, 11.
14. J.A.Schellman, C. R. Trav. Lab. Carlsberg, Ser. Chim.,1955, *29*, 230.
15. S.J. Gill and L.J. Noll, J. Phys. Chem., 1972, *76*, 3065.
16. H.E. Affsprung, S.D. Christian, and J.D. Worley, Spectrochimica Acta, 1964, *20*, 1415.
17. F. Mohamadi, N.G.J. Richards, W.C. Guida, R. Liskamp, M. Lipton, C. Caufield, G. Chang, T. Hendrickson and W.C. Still, J. Comp. Chem., 1990, *11*, 440.
18. A.J. Doig, Ph.D. Thesis, University of Cambridge, 1991.
19. A.R. Fersht, Trends in Biochem. Sci., 1987, *12*, 301.
20. B.A.Shirley, P. Stanssens, U. Hahn, and C.N. Pace, Biochemistry, 1992, *31*, 725.

On the Role of Preorganization in Molecular Recognition

P. A. Kollman

DEPARTMENT OF PHARMACEUTICAL CHEMISTRY, UNIVERSITY OF CALIFORNIA, SAN FRANCISCO, CALIFORNIA 94143-0446, USA

The role of preorganization is one of the two key factors in molecular recognition, the second being interaction complementarity of the preorganized host with the guest. Both of these factors are critical in strong host-guest complementarity. The Nobel Prize in Chemistry in 1987 was given for the important contributions to understanding these two factors by Pederson, Cram and Lehn.

A personal view of this Nobel Prize is that Pederson[1] received it because 18-crown-6, that he first synthesized, was the first example of the group of preorganized hosts, with strong and selective binding of guests. Cram[2] and Lehn[3] elaborated on this theme in diverse and impressive ways. The vision of these two scientists has helped to carry the field a significant distance since its infancy. It is impressive that, despite the complexity of the molecules that they and their groups have made, they have been designed from simple chemical principles, with little or no input from computer modeling techniques. However, it is only the word computer that should be left out in Cram's case, because he is legendary in his ability to manipulate and use the insights from CPK models.

What do computer simulation techniques have to offer this field. Now, in contrast to 10 years ago, the technology has improved to allow the calculation of relative and absolute free energy of association in host-guest interactions in nonpolar and aqueous solutions.[4,5] Structural insights are also derivable from these calculations and they allow one to understand why the free energies come out the way they do in a more fundamental way.

What has changed to allow computer simulation techniques to be applied to this problem? Both changes in computer technology and new algorithmic developments have contributed. However, of equal importance has been the appreciation, based on simulation on pure liquids and extremely preorganized systems, that simple analytical energy functions are surprisingly accurate in reproducing many properties of molecular conformations and interactions. Thus, "molecular mechanics", in which one does not have to consider electrons explicitly but does include electronic structure effects implicitly, can be used in simulations of molecular recognition. However, to sample conformational space, these molecular mechanical energy functions must be combined with Monte Carlo or molecular dynamics methods and then, either of these can be used to calculate free energies by "mutating" one molecular system to another or driving a system along a reaction path.[6] This is feasible because one can calculate the molecular mechanical energy and its derivatives so rapidly, and thus, frequently for large and complex systems, something that cannot be done using *ab initio* quantum mechanical approaches for electronic and nuclear energies. Only Warshel's EVB quantum mechanical approach[7] is comparably rapid to calculate as molecular mechanics; semiempirical MO theory can be used, but is still quite time consuming and, often, significantly less accurate for intermolecular interactions.[8] Thus, the method of choice in studying non-covalent molecular recognition should clearly be molecular mechanics, combined with molecular dynamics or Monte Carlo.

What are the main weaknesses of these methods? There really are two fundamental ones. The first has been alluded to above; the ability of the molecular mechanical potential energy to reproduce the correct (electrons and nuclei) energy of the system can be questioned. Albeit not perfect, suitably parameterized molecular mechanics can be used effectively on a wide variety of systems. But the second, more difficult problem, is the sampling problem; i. e. how to find and determine all the relative free energies of complex molecules in solution. The lowest free energy structures of 18-crown-6 have been determined in the gas phase,[8] but a complete characterization in solution has not been done. What one would ideally like is to characterize the complete free energies for 18-crown-6 and hexaglyme, its open chain analog, in aqueous solution.[9] This would allow us to calculate the "preorganization" free energy directly for a molecular recognition process. We are, as yet unable to do so. However, on model systems such as 18-crown-6 and hexaglyme, such a free energy analysis in solution is just around the corner. For more complex systems, it is not clear how far into the future we need to look in order to allow complete sampling in solution to be accomplished.

We are thus unable to calculate the preorganization free energy quantitatively, but we can, on the other hand, calculate the interaction free energies. Thus, Warshel and co-workers, in their study of enzyme catalysis,[10] have compared the free energies for modeled transition states in an enzyme active site vs the corresponding groups surrounded by water. They have shown, in a number of cases, that the transition state is more stabilized in the enzyme than in water.

The interesting question is, how can this be, given that water has excellent hydrogen bonding groups and flexibility to stabilize either cations and anions in a structurally optimum manner. The answer is that whereas water can have an optimum solute-solvent interaction with polar or ionic solvents, the solvent-solvent energy is raised by about 1/2 the favorable solute-solvent interaction energy. This is as expected from macroscopic electrostatic theory in which one considers the solvent a polarizable medium, where the factor of 1/2 follows from the analytical equations for stabilization of a charge or dipole.

Let us consider a specific example. Potassium ion(K+) has an enthalpy of solvation in water of the order of -70 kcal/mole, which can be separated into -140 kcal/mole of favorable K+ interaction with the water molecules and about +70 kcal/mole increase in the water-water energy when K+ dissolves in it. A typical preorganized ionophore, the calixspherand synthesized by Reinhoudt and co-workers, is only able to achieve about -80 kcal/mole favorable interaction with the K+. However,[11] the ionophore is preorganized, so that its internal energy only goes up by a few kcal/mole when the K+ binds. Thus, the net energy and free energy favors K+ binding to the ionophore, despite the much poorer alignment of the calixspherand than water dipoles in the first shell of the ion. Thus, one cannot use simply the structure of the ionophore to infer its ability to complex ions, since a better ion-ligand interaction can be obviated by the requirement for a greater reorganization energy.

Recently, Miyamoto and Kollman demonstrated the same principle holds for Van der Waals interactions by studying the absolute free energy of association of biotin-avidin,[12] the strongest known small molecule- protein complex, with a binding constant of approximately 10**15. The stabilization of this complex is due to the exquisitely preorganized cavity, which enables 14-18 kcal/mole more Van der Waals attraction to be realized by biotin in the protein binding site than in water.

How can this be? To fully understand it, we must turn to the relative solubility in water of small hydrocarbons. The absolute free energy of solvation of methane in water is about 2.0 kcal/mole,[13] using a 1M standard state. Ethane is actually -0.2 kcal/mole more soluble than methane and propane is only 0.2 kcal/mole less soluble than ethane. These interactions between hydrocarbon and water are almost entirely Van der Waals and the positive free energy for methane comes from the repulsive part of the Van der Waals function.[14] The solvation free energy difference between methane, ethane and propane is so small because the repulsive and attractive Van der Waals contributions come so close to cancelling each other. What is the reason which this much repulsive Van der Waals energy is experienced by these hydrocarbons. It is a fundamental manifestation of the "hydrophobic effect"; water molecules in the first shell of the hydrocarbon attempt to retain optimum hydrogen bonding with each other and are willing to pay a small Van der Waals repulsion price in order to do so. Water is unique in the size of the Van der Waals repulsion contribution to the free energy of solvation, as shown by Rao and Singh.[15] In that study, they compared the free energy of growing methyl groups on solutes in water, methanol and DMSO. Only in water was the initial free energy repulsive, before the attractive dispersion contribution took over.

When an optimum binding site is created in the protein, as in avidin, preorganization allows a well shaped binding site where the ligand can experience only Van der Waals attraction, with little exchange repulsion. Thus, preorganization is a key in Van der Waals interactions as well as electrostatic, but for different reasons. Preorganization allows adequate (not as good as) dipolar alignment without paying the unfavorable solvent-solvent price that water does. Preorganization allows favorable Van der Waals dispersion, without paying the Van der Waals repulsion price that is experienced by the solute in water because of the driving force for water to retain its hydrogen bonded network.

REFERENCES

1. C.J. Pederson, Science, 1988, 240, 760; J Incl. Phenom., 1988, 6, 337.
2. J.M. Lehn, Chem Scr., 1988, 28, 237; J. Incl. Phenom., 1988, 6, 351.
3. D.J. Cram, Science, 1988, 241, 536; J. Incl. Phenom., 1988, 6, 397.
4. W. Jorgensen, Acc. Chem. Res., 1989, 22, 184.
5. P. Kollman and K. Merz, Acc. Chem. Res., 1990, 23, 246.
6. D.L. Beveridge and F.M. diCapua, Ann. Rev. Biophys. Biophys. Chem., 1989, 18, 431.
7. A. Warshel and R. Weiss, J. Amer. Chem. Soc., 1980, 102, 6218.
8. Y. Sun and P. Kollman, J. Comp. Chem., 1992, 13, 33.
9. G. Wipff, P. Weiner and P. Kollman, J. Amer. Chem. Soc., 1982, 104, 3249.
10. A. Warshel and J. Aqvist, Ann. Rev. Biophys. Biophys. Chem. , 1991, 20, 269.

11. S. Miyamoto and P. Kollman, "Molecular Dynamics Studies of Calixspherand Complexes with Alkali Cations: Calculation of the Absolute and Relative Free Energies of Binding of Cations to a Calixspherand", J. Amer. Chem. Soc. (in press).

12. S. Miyamoto and P. Kollman, "Absolute and Relative Binding Free Energy Calculations of the Interaction of Biotin and its Analogs with Streptavidin Using Molecular Dynamics/Free Energy Calculations" J. Amer. Chem. Soc., submitted.

13. A. Ben Naim and Y. Marcus, J. Chem. Phys., 1984, 81, 2016.

14. P.A. Bash, U.C., Singh, R. Langridge and P. Kollman, Science, 1987, 236, 564.

15. B.G. Rao and U.C. Singh, J. Amer. Chem. Soc., 1989, 111, 3125; 1990, 112, 3803.

Self-assembly Through Networks of Hydrogen Bonds

J. P. Mathias, C. T. Seto, J. A. Zerkowski, and
G. M. Whitesides

DEPARTMENT OF CHEMISTRY, HARVARD UNIVERSITY, CAMBRIDGE MA 02138, USA

1. INTRODUCTION

We are developing self-assembly as a strategy for the
preparation of large supramolecular structures. This
approach to synthesis focuses on forming networks of weak,
reversible non-covalent interactions between the
constituent molecules to generate a thermodynamically
stable structure, rather than on forming individually
strong, covalent bonds. A major intellectual impetus for
the development of self-assembly in organic chemistry is
the range of self-assembled structures found in living
organisms.[1] Pertinent examples of these structures
include (i) [t]RNA molecules, whose structures are dictated
by series of hydrogen bonds between A:U and G:C residues,[2]
and (ii) telomers, whose structures are stabilized by
stacked, cyclic hydrogen bond networks between G
residues.[3]

The objective of our research is the preparation of
series of related self-assembled structures, using
hydrogen bonds as the basic intermolecular links. We
intend to use these series to address certain issues
crucial to the application of self-assembly as a synthetic
strategy.[4-8] These issues include:

An Understanding of Thermodynamics. Enthalpy forms the
thermodynamic basis of covalent synthesis. As a
consequence, the intuition of chemists concerning the
structure, enthalpy, and stability is strong. The
corresponding intuition regarding the relations involving
entropy is generally weaker. In self-assembly,
understanding the interplay between enthalpy and entropy
is central to molecular design. We wish to understand
this interplay in sufficient detail to predict the
stability of new structures.

Considerations in Design. Preorganization of constituent
molecules has emerged as a key consideration in the design

"crinkled'

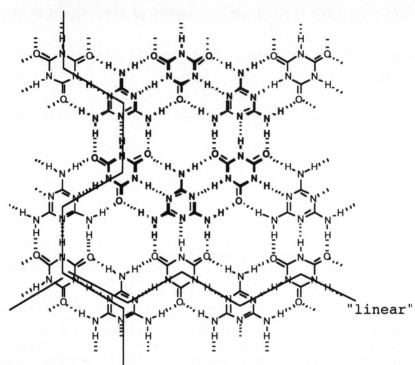

"linear"

<u>Figure 1</u> The putative structure of the CA·M lattice that
is believed to form on mixing equimolar ratios of cyanuric
acid (CA) and melamine (M). The motifs we have extracted
are (i) the cyclic CA_3M_3 unit (bold), (ii) the linear tape,
and (iii) the crinkled tape.

of self-assembling structures. We wish to ascertain where
the balance lies between the rigidity/flexibility of
constituent molecules, the ease of their synthesis, and
the stabilities of complexes derived from them.

Techniques for Characterization. The characterization of
large molecular structures is difficult. We wish to
determine which analytical methods provide the most useful
data for the characterization of non-covalently bound
supramolecular aggregates in organic solution.

    2. SOLUBLE HYDROGEN-BONDED COMPLEXES BASED ON THE
CA·M LATTICE

The strategy we have adopted for the formation of
hydrogen-bonded supramolecular aggregates in organic
solution is based on the synthesis of discrete, soluble
portions of the hydrogen-bonded lattice between cyanuric
acid and melamine (CA·M) (Figure 1).[9] The motif we have

used to build self-assembling structures is the cyclic CA_3M_3 unit (bold) that characterizes this lattice. The CA·M lattice fulfills two key requirements to serve as the basis of self-assembling structures. First, the high density of hydrogen bonds in CA·M provides a strong enthalpic driving force for self-assembly. Second, the cyclic CA_3M_3 unit is a highly symmetrical motif. This factor aids characterization of these supramolecular aggregates, especially by n.m.r. spectroscopy.

Preparation of 2+3 Supramolecular Aggregates.

Mixing equimolar portions of isocyanurates and melamines in chloroform at room temperature results in their association to form random oligomers, with no evidence for the existence of a stable cyclic CA_3M_3 unit in solution. To reduce the unfavorable translational entropy that is associated with self-assembly, and to reduce the loss of conformational entropy upon complexation (by attempting to impose the bound conformation on the unbound constituents), we have used a 'spoke' and 'hub' architecture to preorganize both melamine and isocyanurate units prior to binding.[10] This approach to self-assembly is illustrated in Scheme 1. Two equivalents of the trivalent melamine derivatives, hubM3 and its more flexible analog trisM3, self-assemble with three equivalents of the dimeric isocyanurate derivatives, benzCA2 and furanCA2, to afford 2+3 adducts in chloroform solution.[11] These complexes are stabilized by 36 hydrogen bonds between the five constituent molecules. We have used five analytical techniques to characterize these 2+3 supramolecular aggregates.

1. Solubility: The solubility of the bisisocyanurate derivatives in chloroform is low (<0.1 mM). On complexation with hubM3, the bisisocyanurates dissolve. The ability of two equivalents of hubM3 to solubilize three equivalents of benzCA2 provides qualitative evidence for the stoichiometry of the resulting complex.

2. UV Spectroscopy: The changes in λmax that occur on addition of benzCA2 to hubM3, up to the stoichiometry required for formation of the 2+3 complex, provide quantitative evidence for the stoichiometry of the complex. Beyond this point, no further changes occur.

3. [1]H N.M.R. Spectroscopy: The solubility of these complexes in chloroform has allowed their structural characterization by [1]H n.m.r. spectroscopy. Figure 2 shows the changes that occur in the [1]H n.m.r spectrum in chloroform on going from uncomplexed hubM3 to (hubM3)2(benzCA2)3. The spectrum of hubM3 is broad as a result of self-association and of hindered rotation about

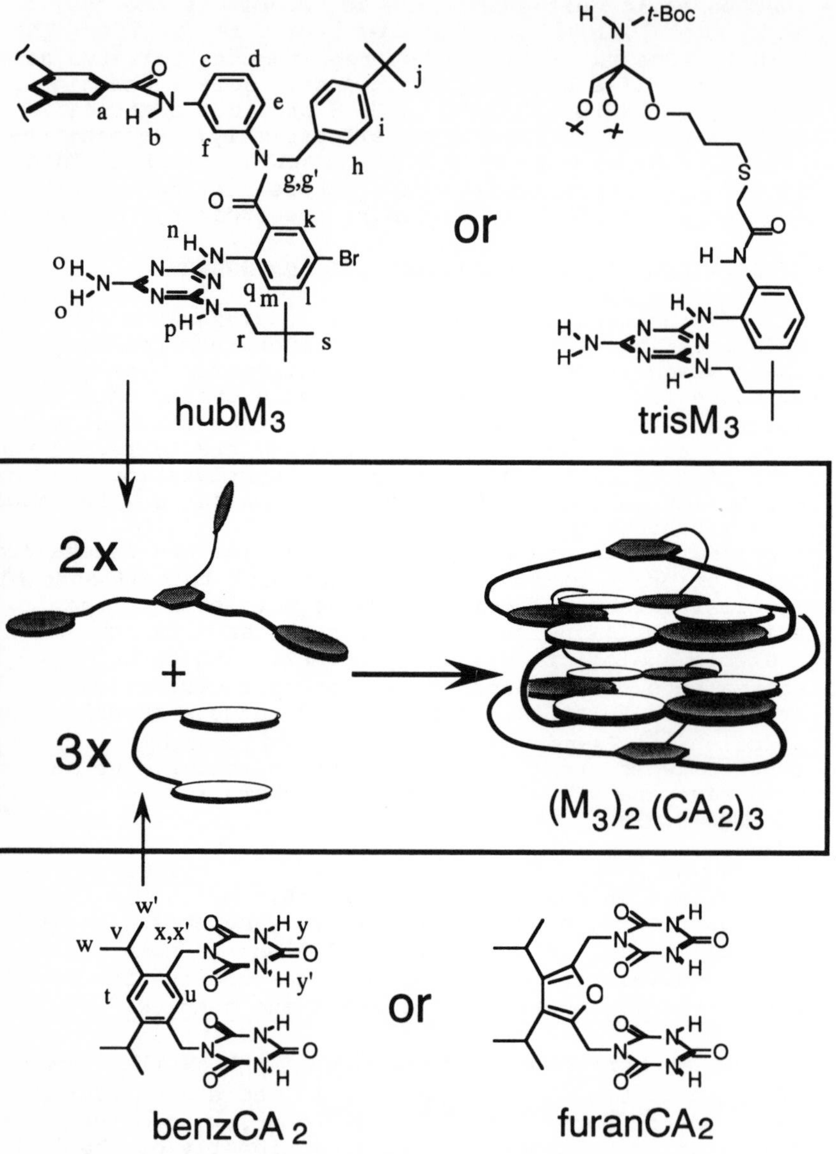

Scheme 1 Self-assembly of the trivalent melamine derivatives, hubM₃ and trisM₃, with the bivalent isocyanurate derivatives, benzCA₂ and furanCA₂, to form 2+3 supramolecular aggregates. The molecular weight of (hubM₃)₂(benzCA₂)₃ is 5.519 KDa and the molecular weight of (trisM₃)₂(benzCA₂)₃ is 4.266 KDa.

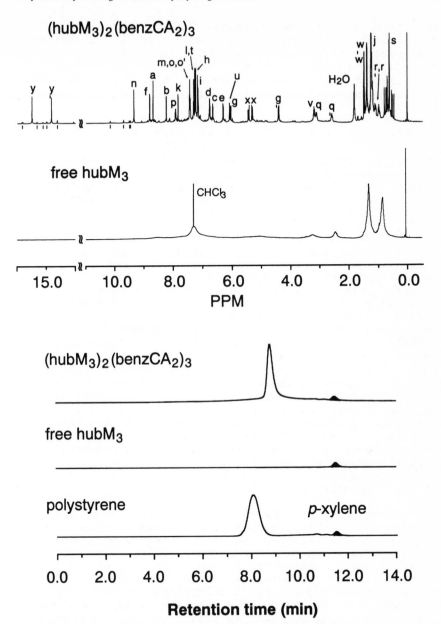

Figure 2 Characterization of (hubM₃)₂(benzCA₂)₃ by ¹H n.m.r spectroscopy and gel permeation chromatography. The annotation on the top ¹H n.m.r. spectrum corresponds with that for (hubM₃)₂(benzCA₂)₃ shown in Scheme 1. The retention time in g.p.c. increases as the size of the molecle decreases. The polystyrene standard has a mean molecular weight of 5050 and a polydispersity of 1.05.

the amide and triazine-NH bonds. In contrast, the
resonances in the spectrum of $(hubM_3)_2(benzCA_2)_3$ are sharp
and predominantly (>95%) those of a single species. These
resonances are consistent with the proposed structure of
$(hubM_3)_2(benzCA_2)_3$. At intermediate stoichiometries, the
spectra show only resonances associated with
$(hubM_3)_2(benzCA_2)_3$ and uncomplexed $hubM_3$. This
observation is consistent with the operation of
cooperativity in the formation of $(hubM_3)_2(benzCA_2)_3$. In
contrast to $(hubM_3)_2(benzCA_2)_3$, the more flexible
$(trisM_3)_2(benzCA_2)_3$ aggregate exhibits considerable
conformational isomerism. Full structural assignment of
this complex is, therefore, impossible. This feature
illustrates the importance of preparing supramolecular
aggregates that are highly structured and symmetrical.

4.   Gel Permeation Chromatography (g.p.c.): G.p.c.
is a technique that separates molecules on the basis of
their hydrodynamic radii. It is a useful technique for
analyzing non-covalently bound structures in organic
solution, since it provides information (albeit
qualitative information) about both stability and size.
Figure 2 shows the g.p.c. traces of uncomplexed $hubM_3$,
$(hubM_3)_2(benzCA_2)_3$, and a polystyrene standard in
chloroform. Uncomplexed $hubM_3$ exhibits self-association
in solution and exists as a poorly-defined mixture of
aggregates. The g.p.c. trace of $hubM_3$ is very broad and
featureless. In contrast, the g.p.c trace of
$(hubM_3)_2(benzCA_2)_3$ shows a single, sharp peak that is
consistent with the presence of a single species in
solution. The profile of the peak suggests that the
stability of the complex is high, with no evidence of
decomplexation over the duration of the run. We believe
that the shapes of peaks for different hydrogen bonded
aggregates in g.p.c. provide a useful qualitative measure
of their relative stabilities.

5.   Vapor Pressure Osmometry (v.p.o.): V.p.o. is a
technique for obtaining molecular weights of molecules in
solution, by correlating the vapor pressure of a solution
containing a compound of unknown molecular weight to that
of solutions of standards of known molecular weights. We
have determined the molecular weights of our 2+3
supramolecular aggregates in chloroform solution, to an
accuracy of ± 10%, by correlating results against those of
N,N' bis[t]Boc-gramicidin-S (MW 1342), sucrose octaacetate
(MW 679), polystyrene (MW 5050), and perbenzoyl β-
cyclodextrin (MW 3321). We observe a significant range of
values for molecular weights depending on the standard
used. We are investigating the effects that self
association and solvent association, caused by hydrogen
bond sites on the periphery of our molecules, have on the
structures in solution (and, therefore, on the inferred
molecular weights) of these complexes. An improved

understanding of these features should allow us to apply
v.p.o. to the analysis of supramolecular aggregates with
greater accuracy.

An Assessment of the Thermodynamic Parameters of Self-Assembly.

To improve our understanding of the thermodynamics of
self-assembly, we have examined the rates of exchange of
components in self-assembled complexes with analogous
structures, uncomplexed in solution, by ^{1}H n.m.r
spectroscopy (Scheme 2). The 1+1 complexes formed between
the trivalent melamine derivatives M_3 and the
complementary trivalent isocyanurate derivatives CA_3 are
the most stable we have prepared. By measuring the rate
constant for exchange at various temperatures, we can
obtain the thermodynamic parameters that are associated
with the transition state of the exchange process. These
values are given in Scheme 2. The exchange process is
unimolecular. This observation suggests that the
transition state for exchange corresponds to the fully
decomplexed state. Dividing the enthalpy of activation
for this exchange (*i.e.* for the dissociation reaction)
(24 kcal mol^{-1}) by the number of hydrogen bonds (18) gives
an individual hydrogen bond enthalpy of 1.3 kcal mol^{-1}.
Although this analysis contains some serious and debatable
assumptions, the value of the hydrogen bond strength
compares favorably with values obtained by other
authors.[12-14]

Qualitative Predictions of Stability.

Based on the enthalpy for each hydrogen bond from the
previous section, we have constructed a semi-quantitative
analysis of the relative contributions of different free
energy terms to self-assembly processes. An example,
based on the assembly of (hubM$_3$)$_2$(benzCA$_2$)$_3$, is shown in
Scheme 3. The enthalpy of formation of 36 hydrogen bonds
in the two cyclic CA$_3$M$_3$ units should be ~48 kcal mol^{-1}.
The term for translational entropy represents the loss in
entropy on bringing the five molecules together in space.
The term for rotational entropy governs the loss in
entropy on bringing together the binding regions of the
melamine and isocyanurate pieces. The term for
conformational entropy relates to the loss in entropy on
restricting rotation about the flexible bonds in hubM$_3$.
We estimate that each arm of hubM$_3$ contains 7 bonds that
must be frozen into one of two rotamers. The change in
conformational entropy on constraining two hubM$_3$ molecules
is, therefore, 2 X (21 x $-RT\ln(1/2)$) = -17 kcal mol^{-1}.
BenzCA$_2$ has no rotamers that are restricted as a
consequence of binding. Combining these values leads to a
predicted free energy of formation of ~ -13 kcal mol^{-1} for
(hubM$_3$)$_2$(benzCA$_2$)$_3$.

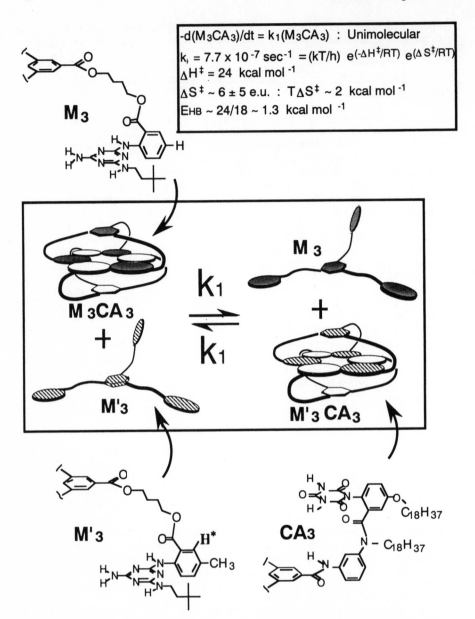

$-d(M_3CA_3)/dt = k_1(M_3CA_3)$: Unimolecular

$k_1 = 7.7 \times 10^{-7} \text{ sec}^{-1} = (kT/h) \, e^{(-\Delta H^\ddagger/RT)} \, e^{(\Delta S^\ddagger/RT)}$

$\Delta H^\ddagger = 24$ kcal mol^{-1}

$\Delta S^\ddagger \sim 6 \pm 5$ e.u. : $T\Delta S^\ddagger \sim 2$ kcal mol^{-1}

$E_{HB} \sim 24/18 \sim 1.3$ kcal mol^{-1}

M$_3$

M$_3$CA$_3$ **k$_1$**   **M$_3$**
   +
M'$_3$
M'$_3$ CA$_3$

M'$_3$ **CA$_3$**

Scheme 2 The thermodynamic parameters associated with the transition state for the exchange process illustrated. The reaction was followed by monitoring the signal of the proton indicated on M'$_3$ by an asterisk by ^{1}H n.m.r. spectroscopy.

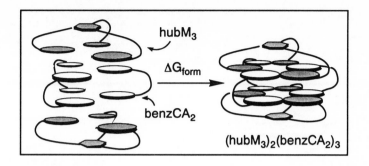

Enthalpy of Formation : $\Delta H_{formn} \sim -48$ kcal mol^{-1}

Translational Entropy : $-T\Delta S_{trans} \sim +12$ kcal mol^{-1}

Rotational Entropy : $-T\Delta S_{rotn} \sim + 6$ kcal mol^{-1}

Conformational Entropy : $-T\Delta S_{confn} \sim +17$ kcal mol^{-1}

$$\Delta G_{form} \sim -48 + 35 \sim \textbf{-13 kcal mol}^{-1}$$

Scheme 3 Semiquantitative estimates of the contributions of enthalpy and entropy to the free energy of self-assembly for $(hubM_3)_2(benzCA_2)_3$.

A second qualitative predictor that we have applied to our systems is the number HB/(N-1). HB is the number of hydrogen bonds in the aggregate. N is the number of constituent molecules in the aggregate. The larger the value of HB, the more stable the aggregate; the larger the value of N, the less stable the aggregate. The ratio HB/(N-1) is not correlated in a fundamental way to ΔG, but HB correlates with the enthalpy of formation of the network of hydrogen bonds, and (N-1) correlates with the loss of translational entropy on formation of the aggregate. Large values of HB/(N-1) should, therefore, correlate with high stability for these types of complexes. Figure 3 shows the values of HB/(N-1) for six types of our supramolecular aggregates. These structures range from the untethered cyclic CA_3M_3 unit to the very stable 1+1 complex. The trend in stabilities that is suggested in this figure matches that we have observed.

Although these methods for assessing the thermodynamic aspects of self assembly are both semi-quantitative, they do represent useful methods of

HB/(N-1)

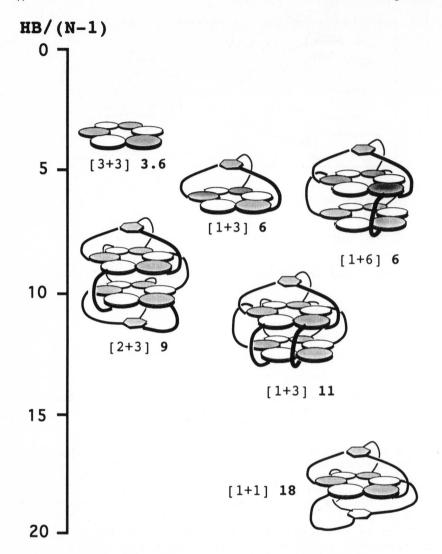

Figure 3 Values of HB/(N-1) for six self-assembled supramolecular aggregates. An increase in the value of HB/(N-1) should indicate an increase in the stability of the supramolecular aggregate. The order of these values of HB/(N-1) correlates qualitatively with the stabilities of the complexes. The numbers in brackets refer to the numbers of melamine units (first) and isocyanurate units (second) involved in the supramolecular aggregate.

predicting which systems may self assemble to give stable structures and which systems may have low stability.

3. CO-CRYSTALLIZATION : HYDROGEN-BONDED SELF-ASSEMBLY IN THE SOLID STATE.

The vast literature on hydrogen bonding in crystals is a testament to the power of this interaction as a stabilizing feature in the solid state.[15,16] Crystals that contain extensive networks of hydrogen bonds provide a means to investigate the effects of molecular structure on solid-state packing, and, ultimately, to design and build useful solids based on molecular crystals. This strategy for crystal engineering requires a step beyond the level of molecular design used to prepare soluble molecules, and must be able to tackle problems that are unique to the solid state, such as crystal packing forces.

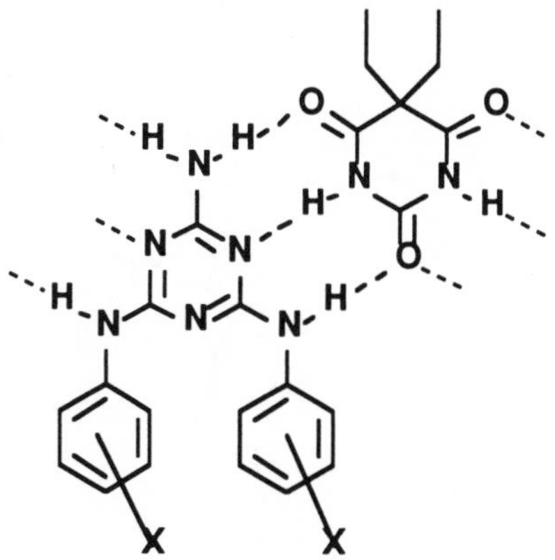

Figure 4 The series of co-crystals we have obtained demonstrates the effect of changing the phenyl substituent (X) on the crystal structure of the 1:1 complex between the substituted diphenylmelamine derivatives and barbital. In all cases, the triad hydrogen bond network between the substituted diphenylmelamines and barbital is retained.

The basis for our efforts in crystal engineering is the collection of motifs found within the CA·M lattice (Figure 1). By using substituted melamines and isocyanurates, we can interfere with the hydrogen bond network and 'excise' portions of the lattice. We have used series of co-crystals formed from equimolar ratios of substituted diarylmelamine derivatives and barbital (Figure 4) to identify three motifs, (i) linear tapes, (ii) crinkled tapes, and (iii) rosettes based upon the cyclic CA3M3 structure.[9] These motifs are shown in Figure 5. They represent a basis set that we are using to probe systematically the interactions that determine the structure of the crystal. We have shown that small changes in the size and position of the substituent on the phenyl ring of the diarylmelamine derivatives can be performed without disrupting the triad pattern of hydrogen

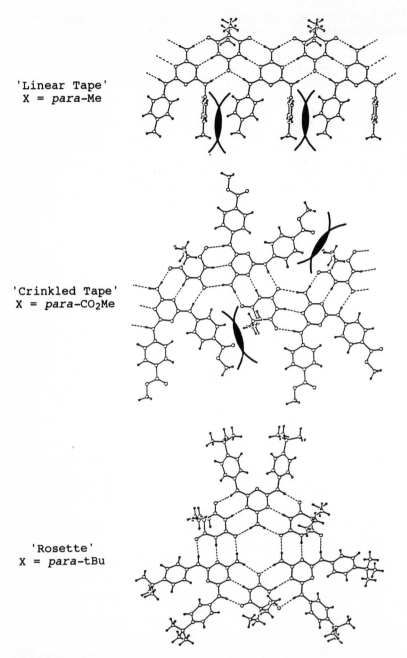

'Linear Tape'
X = *para*-Me

'Crinkled Tape'
X = *para*-CO$_2$Me

'Rosette'
X = *para*-tBu

Figure 5 The effects on the crystal structure of the steric requirements of the *para*-substituents on the diarylmelamine derivatives. We propose that the differentiation of linear tape, crinkled tape, and rosette motifs is determined by the unfavorable steric interactions that are indicated on the structures.

bonds between adjacent pieces.[17] Instead, this substitution influences the motif adopted in the crystal. Co-crystals barbital and *para*-(F, Cl, Br, Me, CF₃) and *meta*-(F, I, Me, CF₃) substituted diarylmelamines adopt the linear tape motif; co-crystals between *para*-CO_2Me-diarylmelamine and barbital adopt the crinkled tape motif; co-crystals between *para*-^{t}Bu-diarylmelamines and barbital adopt the rosette motif. We have used steric arguments to rationalize the interactions that are responsible for the differentiation between the respective motifs. We consider the linear tape to be the lowest energy motif based on its prevalence. The crinkled tape is adopted when steric interactions between adjacent *para*-substituents on the diarylmelamines become unfavorable in the linear tape. These interactions are relieved in the crinkled tape. Further increases in the steric bulk of the *para*-substituents create an unfavorable contact with the proximal ethyl groups of the barbital moieties. In this case, the steric strain is relieved by the adoption of the rosette motif, in which the substituents on the diarylmelamine groups are at their most distant.

4. CONCLUSIONS AND FUTURE PROSPECTS

We have prepared a series of supramolecular aggregates based on networks of hydrogen bonds. These structures demonstrate the versatility of the CA·M lattice as the basis for constructing self-assembling supramolecules. In the solution state, these molecules are characterized by 18 or 36 hydrogen bonds, 4-7 constituent pieces, molecular weights in the range 3-7 KDa, various levels of preorganization, and a range of stabilities. We are now in a position to (i) test certain hypotheses (for example the relationship of HB/(N-1) to stability); (ii) assess the relative stabilities of many of these complexes by ^{1}H n.m.r. exchange experiments; (iii) assess critically our criteria for design (in particular, how effective is our preorganization?); (iv) apply molecular modeling to assist our design and to correlate structural information; (v) prepare structures with increasing stabilities, and consequently; (vi) prepare structures of increasing size. Most importantly, we have obtained a series of closely-related self-assembled structures that is allowing us to begin to decipher trends and make predictions about new supramolecular aggregates.

In the solid state, we have applied an approach to crystal design based upon a closely-related series of structures. The central features of this program are (i) the observation of trends and (ii) the development of a capability for prediction. We believe that we can predict secondary architecture--for example, the occurrence of tapes versus rosettes (Figure 5)--with some confidence, based on steric arguments . We are now tackling two important issues. The first problem is that

of polymorphism, *i.e.* how many crystal phases are
thermodynamically accessible? Which are kinetically
preferred? The use of powder diffraction methods and
crystallization from different solvents should determine
whether the crystal structures we have observed are
determined by kinetic or thermodynamic factors. The
second issue is the control of tertiary architecture. Can
we prepare designed solids by introducing new levels of
orthogonal non-covalent interactions that will control the
crystal packing of our basic tape motifs?

Molecular self-assembly is in its infancy. Any real
application of self-assembly, in either solution or solid
states, will depend on the availability of fundamental
information concerning thermodynamics, stabilities of
complexes, rules for design, and techniques for
characterization. When this fundamental information is
available, it will be possible to design and synthesize
new types of structures efficiently.

ACKNOWLEDGEMENTS

This work was supported by the National Science
Foundation (Grants CHE-88-12709 to GMW, DMR 89-20490 to
the Harvard University Materials Research Laboratory, and
CHE-80-00670 for the purchase of a Siemens X-Ray
Diffractometer). NMR instrumentation was supported by the
National Science Foundation Grant CHE-84-10774. The
Harvard University Chemistry Department Mass Spectrometry
Facility was supported by National Science Foundation
Grant CHE-90-20043 and National Institute of Health Grant
1 S10 RR06716-01. JPM acknowledges support from an
SERC/NATO Postdoctoral Fellowship, 1991-93. CTS was an
Eli Lilly Predoctoral Fellow, 1991. We thank Professor
Robert Cohen (MIT, Chemical Engineering) for loan of the
vapor pressure osmometer.

REFERENCES

1.   J.S. Lindsey, New J. Chem, 1991, 15, 153.
2.   D.E. Draper, Acc. Chem. Res., 1992, 25, 201.
3.   C. Kang, X. Zhang, R. Ratliff, R. Moyzis, and A.
     Rich, Nature, 1992, 350, 126.
4.   G.M. Whitesides, J.P. Mathias, and C.T. Seto,
     Science, 1991, 254, 1312.
5.   J.-M. Lehn, Angew. Chem. Int. Ed. Engl., 1990, 29,
     1304.
6.   P. Tecilla, R.P. Dixon, G. Slobodkin, D.S. Alavi,
     D.H. Waldbeck, and A.D. Hamilton et al., J. Am. Chem.
     Soc., 1990, 112, 9408.
7.   J.F. Stoddart et al., J. Am. Chem. Soc., 1992, 114,
     193.
8.   S.C. Zimmerman and B.F. Duerr, J. Org. Chem., 1992,
     57, 2215.

9. J.A. Zerkowski, C.T. Seto, and G.M. Whitesides, <u>J. Am. Chem. Soc.</u>, 1992, In press.
10. C.T. Seto, and G.M. Whitesides, <u>J. Am. Chem. Soc.</u>, 1990, <u>112</u>, 6409.
11. C.T. Seto, and G.M. Whitesides, <u>J. Am. Chem. Soc.</u>, 1991, <u>113</u>, 712.
12. (a) L.D. Williams, B. Chaula, and B.R. Shaw, <u>Biopolymers</u>, 1987, <u>26</u>, 591.
13. A.R. Fersht, <u>Trends in Biochem. Sci.</u>, 1987, <u>12</u>, 301.
14. H.-J. Schneider, <u>Angew. Chem. Int. Ed. Engl.</u>, 1991, <u>30</u>, 1417.
15. C.R. Desiraju, 'Crystal Engineering: The Design of Organic Solids', Elsevier, Amsterdam, 1989.
16. M.C. Etter, <u>J. Phys. Chem.</u>, 1991, <u>95</u>, 460.
17. J.A. Zerkowski, C.T. Seto, D.A. Wierda, and G.M. Whitesides, <u>J. Am. Chem. Soc.</u>, 1990, <u>112</u>, 9025.

Controlling Self-assembly in Organic Synthesis

Peter R. Ashton, Douglas Philp, Neil Spencer and
J. Fraser Stoddart

SCHOOL OF CHEMISTRY, UNIVERSITY OF BIRMINGHAM, EDGBASTON, BIRMINGHAM
B15 2TT, UK

1 PREAMBLE

The creation of nanometre-scale devices has fascinated and inspired[1] the scientific community for many years. Electronic engineers have approached this problem by employing a so-called 'top down' approach (**Figure 1**) involving the etching of components on to wafers of silicon. In the natural world, there are many examples of functioning devices, *e.g.* ion channels, in the nanometre size range which are constructed by self-assembly processes[2] from relatively simple molecular sub-units. Synthetic chemists, however, have generally been content with creating structures which are in the 0.1 to 1 nm size range by the elegant use of reagent or catalyst control.

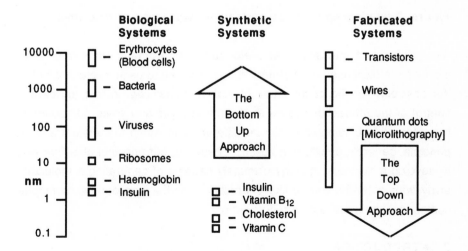

Figure 1 Illustrating the need for nanochemistry (Adapted from Ref. [3]).

We wished to appeal to a new approach to the construction of nanometre-scale devices, namely the so-called 'bottom-up' approach (**Figure 1**), in which small, 'intelligent' wholly-synthetic chemical sub-units guide their own assembly into large, ordered molecular arrays without the need for reagent control or catalysis. The techniques necessary to achieve this goal are present in Nature. The chemist must continue to learn from Nature. The use of relatively weak, non-covalent bonds to organise small synthetic sub-units into the most stable supramolecular arrangements provides the key to molecular self-assembly processes. Therefore, the science of supramolecular[4] or host-guest chemistry[5] is ideally placed to address the problem of self-assembly in organic synthesis. We have recently shown[6-12] that molecular systems, *i.e.* the catenanes and rotaxanes (**Figure 2**), incorporating some degree of mechanical entanglement, may be constructed with comparative ease using self-assembly processes based on the mutual

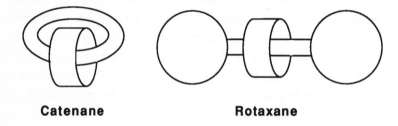

Catenane   **Rotaxane**

Figure 2 Interlocking molecular systems – Catenanes and rotaxanes

recognition and interaction of π-electron rich aromatic units with 4,4'-bipyridinium dications. Furthermore, it has occurred to us that rotaxanes offer the possiblity for the construction of molecular devices, in that, by careful control of the interactions between the macrocyclic and linear components, two alternative binding sites can be differentiated within the system and, in principle, generate a switching action. Therefore, our principal objective was to develop a synthetic strategy ultimately capable of generating a molecular array with potential device-like character from the appropriate molecular components by a process of self-assembly.

2 INTRODUCTION

One of the simplest 4,4'-bipyridinium dications is the 1,1'-dimethyl-4,4'-bipyridinium salt **1**.$2PF_6$ – more commonly known as the herbicide,

Paraquat. This π-electron deficient molecule forms a relatively strong complex[13] with the crown ether, bisparaphenylene-34-crown-10 (BPP34C10), **2** in acetone solution (**Scheme 1**). The complex is stabilised by the electrostatic, π–π stacking, and charge transfer interactions between the π-electron deficient dication and the π-electron rich hydroquinol residues

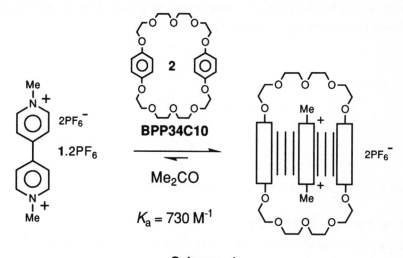

Scheme 1

of the crown ether. There is also some additional stabilisation derived for the interaction of acidic hydrogen atoms – α to nitrogen on the dication – with the ether oxygen atoms of the polyether bridge in the crown ether. The sum of these various interactions amounts to a binding energy of 3.95 kcal mol[-1] in acetone at room temperature. More interestingly, the X-ray crystal structure of this complex reveals (**Figure 3a**) that the dication **1**[2+] is inserted

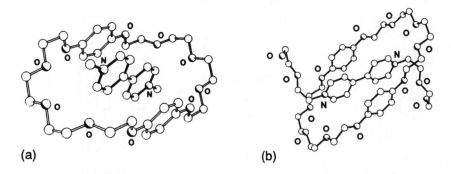

(a)   (b)

Figure 3 The X-ray crystal structure of the complex between (a) Paraquat **1**[2+] and (b) the 1,1'-[bis(2-hydroxyethoxy)ethyl]-4,4'-bipyridinium dication **3**[2+] and BPP34C10 **2**

through the mean plane of the BPP34C10 **2** molecule to give a pseudorotaxane-like[14] structure. The solid state structure of this complex suggested to us that it might be possible to utilise the self-assembly of this pseudorotaxane-like complex in the synthesis of rotaxanes. If we could replace the methyl groups in the dication **1**$^{2+}$ with units amenable to the attachment of blocking groups, then it should be possible to construct a rotaxane by a threading process (**Figure 4**).

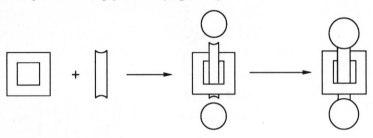

Figure 4 Making a rotaxane by a threading process

3 THE INITIAL STRATEGY

A suitable replacement for the methyl groups in the dication **1**$^{2+}$ was considered to be the (2-hydroxyethoxy)ethyl group. The 4,4'-bipyridinium dication **3**$^{2+}$, carrying this substituent at both nitrogen atoms, was found[15] to form (**Scheme 2**) a strong complex with BPP34C10 **2**. Single crystal X-ray structural analysis of [**2.3**][PF$_6$]$_2$ revealed (**Figure 3b**) that it too has a pseudorotaxane-like structure in the solid state. And so, all that remained

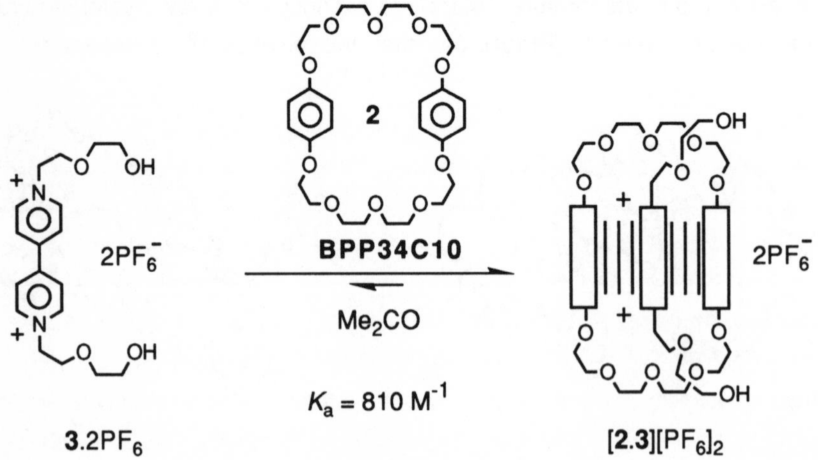

2

BPP34C10

$\xrightarrow{\text{Me}_2\text{CO}}$

$K_a = 810 \text{ M}^{-1}$

2PF$_6^-$

2PF$_6^-$

3.2PF$_6$

[**2.3**][PF$_6$]$_2$

Scheme 2

was for us to identify a suitable blocking group[16] to prevent slippage of the macrocyclic 'wheel' from the 'axle' and to find a synthetic method for attaching the blocking group to the two primary hydroxyl functions at the ends of the 'axle'. The first blocking group we tested was the triphenylmethyl (trityl) group. This group could be easily attached (**Scheme 3**) to the dication 3^{2+} by reaction of the triphenylmethyl hexafluorophosphate with $3.2PF_6$ in the presence of a base at room temperature in acetonitrile, giving the bis(triphenylmethyl) ether $4.2PF_6$ in good yield. Unfortunately, these groups proved to be much too small to prevent the slippage of the macrocycle on and off the 'axle', as evidenced by the immediate development of a strong red colour on addition of BPP34C10 **2** to a solution containing the bis(triphenylmethyl) ether of $4.2PF_6$. This colour is a result of the characteristic charge transfer absorption arising from the interaction between

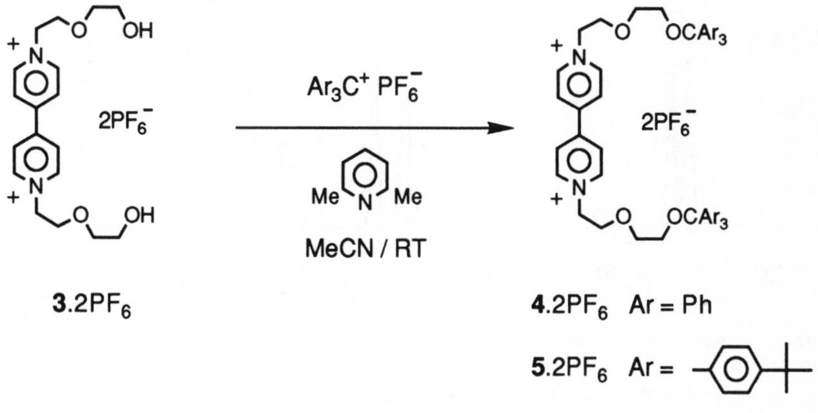

Scheme 3

4,4'-bipyridinium dications and hydroquinol rings. Therefore, an alternative, larger blocking group had to be sought. The tris(4-*t*-butylphenyl)methyl group was identified as a possible candidate for this rôle. Reaction of tris(4-*t*-butylphenyl)methyl) chloride[17] with the dication 3^{2+} in the presence of a base and silver hexafluorophosphate at room temperature in acetonitrile (**Scheme 3**) yielded the bis(tris(4-*t*-butylphenyl)methyl) ether $5.2PF_6$. Addition of BPP34C10 **2** to a solution of $5.2PF_6$ in acetone did not result in the development of the characteristic red colour associated with the interaction of BPP34C10 with a 4,4'-bipyridinium dication. This indicated that the new blocking group was large enough to prevent the crown ether slipping on and off the 'axle'. However, ^1H NMR spectroscopy, unexpectedly,

revealed a problem with this blocking group. The ^{1}H NMR spectrum
(**Figure 5**) of a freshly prepared solution of **5.2PF$_6$** in CD$_3$COCD$_3$ showed
the expected four signals for the *O*-methylene protons of the nitrogen
substituents. On standing, four new signals (❑) developed which may be
assigned to the alcohol and diol produced by the loss of one or both of the
tris(4-*t*-butyl phenyl)methyl) groups from **5.2PF$_6$**, respectively.

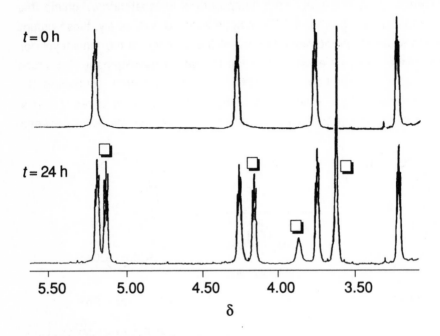

Figure 5 The 400 MHz ^{1}H NMR spectrum of **5.2PF$_6$** in CD$_3$COCD$_3$
recorded at room temperature

The unexpected hydrolytic instability of the tris(4-*t*-butylphenyl)methyl)ether
protecting group precludes its use in the synthesis of rotaxanes. Thus, a new
synthetic strategy was required.

4 A NEW STRATEGY

In devising a new strategy, we wished to take advantage of the fact that,
although 4,4'-bipyridinium dications are strongly bound by BPP34C10 **2**, the
corresponding monoquaternary salts are not bound[18] to any observable
extent. Thus, if we consider what would happen if we react (**Scheme 4**) a
blocking group **BG** containing a leaving group X with 4,4'-bipyridine, then the
first reaction which occurs will produce a monoquaternary salt **MQT.X**. If this

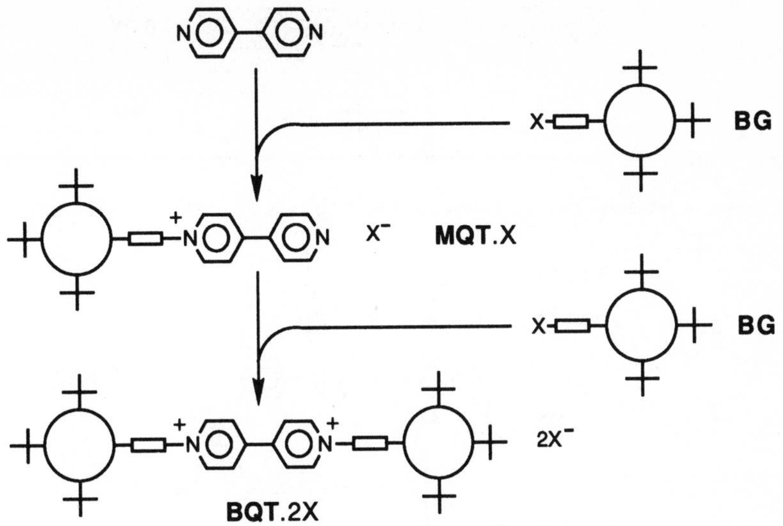

Scheme 4

reaction is performed in the presence of BPP34C10 **2**, then we would **not** expect the crown ether to complex with the **MQT**+ cation. Hence, further reaction of the **MQT**+ cation with **BG** will produce the dumbell shaped dication, **BQT**$^{2+}$, incorporating **no** crown ether. However, if we perform the same reaction (**Scheme 5**), starting from the bismonoquaternary salt **6.2X**, then initial reaction of this salt with **BG** will produce a tricationic species containing both a monoquaternary unit and, more importantly, a 4,4'-bipyridinium dication, capable of complexing with BPP34C10. If this reaction is performed in the presence of **2**, then we would expect the trication

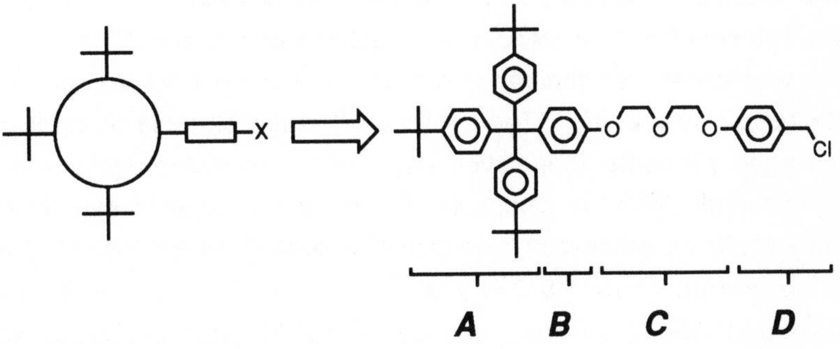

Figure 6 The design of the blocking group **BG**

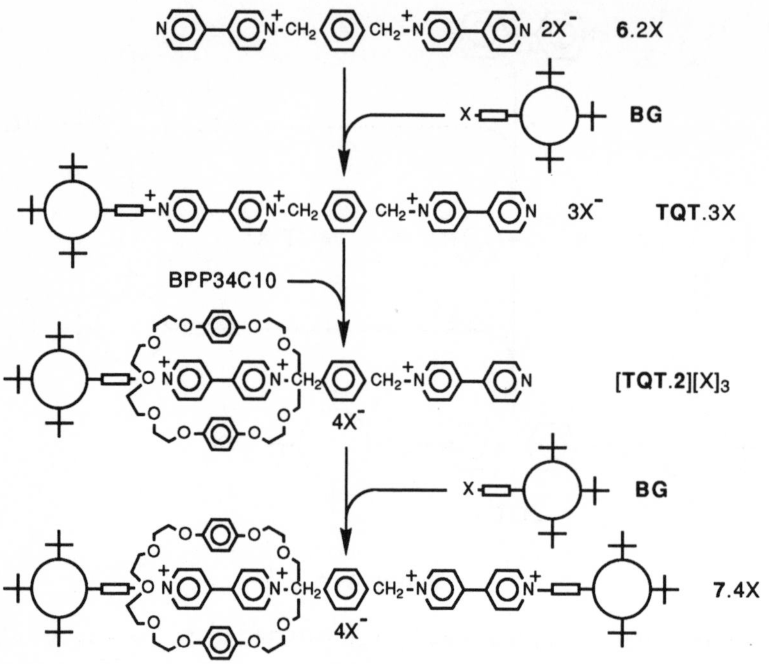

Scheme 5

TQT³⁺ to be bound strongly by the crown ether to form the complex [**TQT.2**][X]$_3$. Further reaction of the tricationic complex [**TQT.2**][X]$_3$ with the blocking group **BG** will produce the [2]rotaxane **7.4X**. In order to enable us to pursue this synthetic strategy it is necessary to design a suitable blocking group **BG**. The salient features of the design are summarised in **Figure 6**. The original tris(4-*t*-butylphenyl)methyl blocking group (**A**) is retained, but in order to improve its stability, an aromatic spacer **B** is added. The spacer polyether chain **C** is incorporated so that the reactive function **D** is located at a position remote from the steric bulk associated with **A** and **B**. A benzylic chloride was chosen as the reactive function **D** since they are known to alkylate bipyridines readily. The synthesis[19] of the blocking group **8** was accomplished (**Scheme 6**) in seven steps from 4-*t*-butylbromobenzene in 39% yield overall. With this compound in hand, the proposed regime of self-assembly could be assessed. Reaction (**Scheme 7**) of BPP34C10 **2**, the bismonoquaternary salt **6.2PF₆** and the benzylic chloride **8** in dry dimethylformamide at 10 kbars pressure[20] for 36 hours produced, after chromatography, the expected [2]rotaxane **7.4PF₆** as a glassy orange solid in 23% yield after counterion exchange. The [2]rotaxane was initially characterised[19] by fast atom bombardment mass spectrometry. The

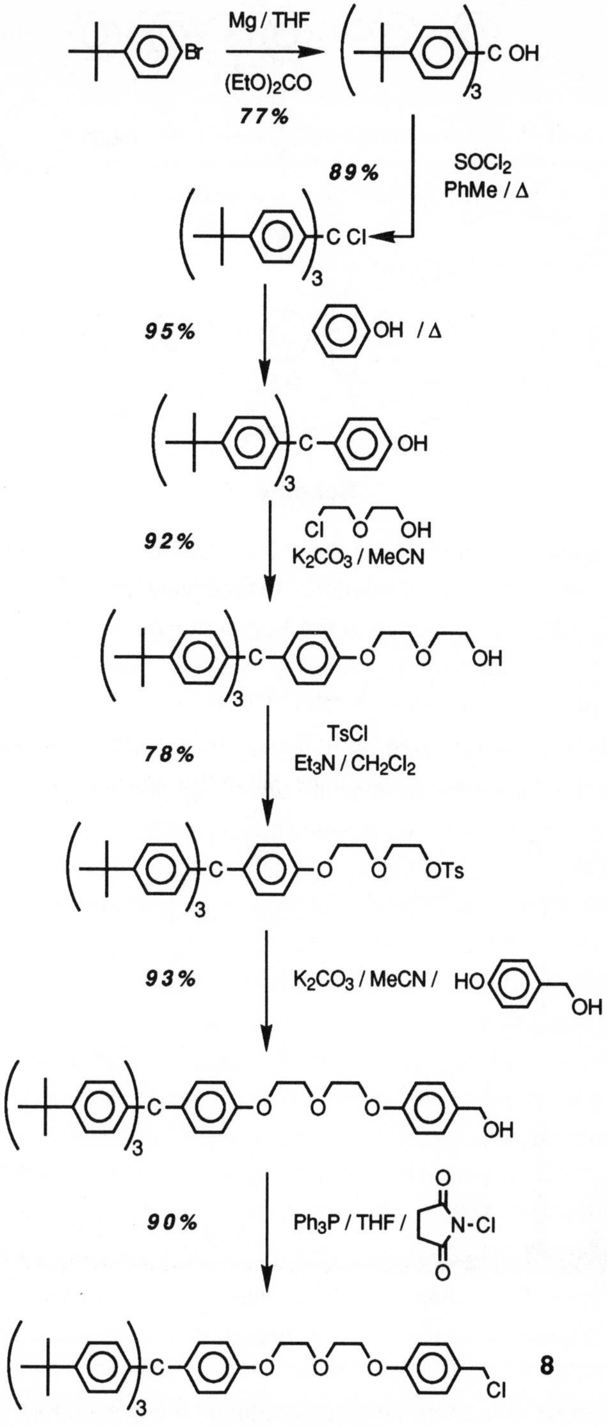
Mg / THF
(EtO)$_2$CO
77%
C OH
3
89%
SOCl$_2$
PhMe / Δ
C Cl
95%
OH / Δ
C
OH
92%
Cl O OH
K$_2$CO$_3$ / MeCN
78%
TsCl
Et$_3$N / CH$_2$Cl$_2$
OTs
93%
K$_2$CO$_3$ / MeCN / HO OH
90%
Ph$_3$P / THF / N-Cl
Cl

8

Scheme 6

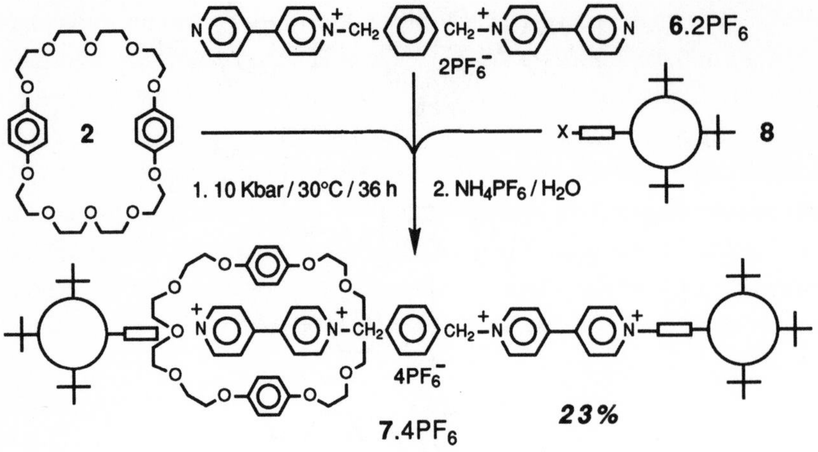

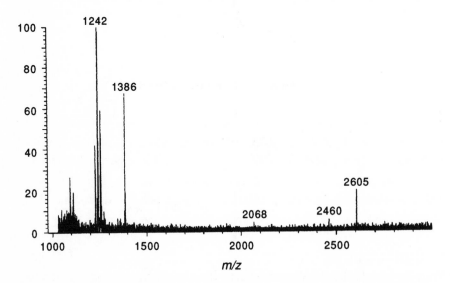

Scheme 7

spectrum revealed (**Figure 7**) peaks at *m/z* 2605 and 2460 corresponding to the loss of two and three counterions, respectively, from **7.4PF$_6$**. The small peak at *m/z* 2068 corresponds to the fragmentation and subsequent loss of the BPP34C10 macrocycle from the 'axle'. Both ^{1}H and ^{13}C NMR spectroscopies indicated that, in CD$_3$COCD$_3$ solution at room temperature, the BPP34C10 macrocycle is moving very rapidly between the two equivalent 4,4'-bipyridinium dications within the tetracationic component of

Figure 7 Positive-ion fast atom bombardment mass spectrum of the [2]rotaxane **7.4PF$_6$**

7.4PF$_6$. The rapid equilibration between the two degenerate translational isomers[21] of the [2]rotaxane **7.4PF$_6$** (**Figure 8**) is indicated by the presence of only two doublets at δ 9.31 and δ 9.27 in the 400 MHz ^{1}H NMR spectrum for the protons α to nitrogen in the tetracationic component of **7.4PF$_6$**. Further evidence came from the presence of only two resonances in the 75 MHz ^{13}C NMR spectrum, at δ 31.5 and δ 34.7, for the methyl and quaternary carbon atoms in the *t*-butyl groups on the stoppers terminating the tetracationic component of **7.4PF$_6$**. This rapid equilibration between the translational isomers means that, in effect, the system is behaving[22] like a molecular shuttle.

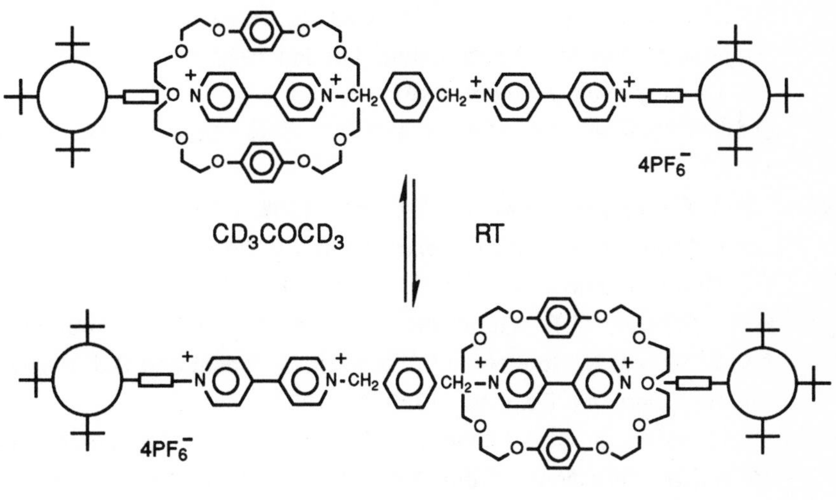

Figure 8 Degenerate translational isomerism in **7.4PF$_6$**

In order to test the selectivity of this self-assembly process, we performed the reaction between 4,4'-bipyridine and the benzylic chloride **8** shown in **Scheme 4** in the presence of BPP34C10 **2**. The outcome of this reaction was as expected, *i.e.* the only product detected was the 4,4'-bipyridinium salt **BQT.2X**, incorporating **no** BPP34C10 **2**. This observation demonstrates the complete selectivity of this self-assembly process.

5 CONCLUSIONS

With the self-assembly of the molecular shuttle **7^{4+}**, we have shown that it is possible to create an apparently complex molecular structure with remarkable ease. In this case, we may regard the precursor components of the molecular structure, namely BPP34C10 **2**, the bismonoquaternary dication 6^{2+} and the benzylic chloride **8**, as intelligent, in that they hold all of the information necessary to guide accurately the construction of 7^{4+} under the appropriate conditions. There is no need for external template control or catalysis. The fact that a molecular structure of nanometre-scale can be constructed with such apparent ease and efficiency bodes well for the construction of other molecular structures and superstructures with addressable functions using a similar synthetic approach based on self-assembly.

References and Footnotes

[1]   R.P. Feynman *Sat. Rev.* **1960**, *432*, 45. See also: K.E. Drexler *'Engines of Creation'* Fourth Estate, London 1990.

[2]   J.S. Lindsey *New J. Chem.* **1991**, *15*, 153.

[3]   G.M. Whitesides, J.P. Mathias and C.T. Seto *Science* **1991**, *254*, 1312.

[4]   J.M. Lehn *Angew. Chem. Int. Ed. Engl.* **1988**, *27*, 89. See also: J.M. Lehn *Angew. Chem. Int. Ed. Engl.* **1990**, *29*, 1304.

[5]   D.J. Cram *Angew. Chem. Int. Ed. Engl.* **1988**, *27*, 1009.

[6]   P.R. Ashton, T.T. Goodnow, A.E. Kaifer, M.V. Reddington, A.M.Z. Slawin, N. Spencer, J.F. Stoddart, C. Vicent and D.J. Williams *Angew. Chem. Int. Ed. Engl.* **1989**, *28*, 1396.

[7]   J.F. Stoddart, In Ciba Foundation Symposium No. 158, *'Host-Guest Molecular Interactions - From Chemistry to Biology'* Wiley, Chichester, **1991**, 5.

[8]   D. Philp and J.F. Stoddart *Synlett* **1991**, 445.

[9]   P.L. Anelli, P.R. Ashton, A.M.Z. Slawin, N. Spencer, J.F. Stoddart and D.J. Williams *Angew. Chem. Int. Ed. Engl.* **1991**, *30*, 1036.

[10]   P.R. Ashton, C.L. Brown, E.J.T. Chrystal, T.T. Goodnow, A.E. Kaifer, K.P. Parry, A.M.Z. Slawin, N. Spencer, J.F. Stoddart and D.J. Williams *Angew. Chem. Int. Ed. Engl.* **1991**, *30*, 1039.

[11]   P.R. Ashton, C.L. Brown, E.J.T. Chrystal, K.P. Parry, M. Pietraszkiewicz, N. Spencer and J.F. Stoddart *Angew. Chem. Int. Ed. Engl.* **1991**, *30*, 1042.

[12]   J.F. Stoddart *Chem. Br.* **1991**, *27*, 714.

[13]   B.L. Allwood, H. Shahriari-Zavareh, N. Spencer, J.F. Stoddart and D.J. Williams *J. Chem. Soc., Chem. Commun.* **1987**, 1064.

[14]   The name *rotaxane* derives (G. Schill *'Catenanes, Rotaxanes and Knots'* Academic Press, New York, 1971) from the Latin *rota* meaning wheel and *axis* meaning axle. The addition of the prefix *pseudo* indicates that the wheel is free to dissociate from the 'axle' as in a more conventional type of complex. In chemical terms, this means that the molecular components of the *pseudorotaxane* are held together only by their mutual noncovalent bonding attraction. See: P.R. Ashton, D. Philp, N. Spencer and J.F. Stoddart *J. Chem. Soc., Chem. Commun.* **1991**, 1677.

[15]   P.R. Ashton, D. Phlip, M.V. Reddington, A.M.Z. Slawin, N. Spencer, J.F. Stoddart and D.J. Williams *J. Chem. Soc., Chem. Commun.* **1991**, 1680.

[16]   The large (10.6 Å by 4.7 Å) and flexible cavity of the BPP34C10 macrocycle dictates that the blocking group chosen must be very large. The use of smaller crown ethers, such as BPP31C9, was discounted because of the very much smaller binding constant observed between BPP31C9 and **1.**2PF$_6$.

[17]   Tris(4-*t*-butylphenyl)methyl chloride is easily prepared in two steps by the action of 4-*t*-butylphenylmagnesium bromide on diethyl carbonate and subsequent reaction of the product so obtained with thionyl chloride in refluxing toluene.

[18]   P.L. Anelli, P.R. Ashton, R, Ballardini, V. Balzani, M. Delgado, M.T. Gandolfi, T.T. Goodnow, A.E. Kaifer, D. Philp, M. Pietraszkiewicz, L. Prodi, M.V. Reddington, A.M.Z. Slawin, N. Spencer, J.F. Stoddart, C. Vicent and D.J. Williams *J. Am. Chem. Soc.* **1992**, *114*, 193.

[19]   P.R. Ashton, D. Philp, N. Spencer and J.F. Stoddart *J. Chem. Soc., Chem. Commun.* Submitted.

[20]   For a comprehensive review of the effect of high pressure on reactions and its use in synthesis, see: N.S. Isaacs *Tetrahedron* **1991**, *47*, 8463.

[21]   Translational isomerism characterises systems in which one molecular component can occupy two (or more) sites within the molecular structure of the noncovalently bonded assembly. In the rotaxane **7.**4PF$_6$, the two sites are degenerate and hence the two translational isomers cannot be distinguished physically or chemically even at low temperatures.

[22]   P.L. Anelli, N. Spencer and J.F Stoddart *J. Am. Chem. Soc.* **1991**, *113*, 5131.

Recognition, Replication and Assembly

Julius Rebek, Jr.

DEPARTMENT OF CHEMISTRY, MASSACHUSETTS INSTITUTE OF TECHNOLOGY,
CAMBRIDGE, MA 02139, USA

'guest' and 'host' are words that intertwine. Converging, mixing, reciprocating...
Guests bring ideas from outside.

Mao II
Don Delillo, (1991)

For some time we have been concerned with the relationship between
molecular recognition and catalysis: how weak intermolecular forces, applied across
complementary structures can lead to rate enhancements[1]. This is little more than an
attempt to realize the Pauling principle-recognizing transition states rather than ground
states-in simple models for enzymes. Here we discuss harnessing these forces for a
unique form of catalysis, *autocatalysis*, and suggest that molecular recognition
phenomena lead inevitably to replication.

For reasons that now unaccountably fail to come to mind, we chose adenine
recognition as the recognition target, and have expended considerable effort in
designing reciprocal surfaces for its purine nucleus[2]. Initially, we availed ourselves
of the unusual structural features of the Kemp triacid to dissect hydrogen-bonding and
aromatic stacking interactions for adenines. The U-shaped relationship between the
carboxyl functions of this module permit the construction of molecules that fold back
upon themselves and the rigidity conferred by the triacid in conjuction with various
aromatic spacers permits the positioning of hydrogen-bonding elements with some
accuracy. (Fig. 1)

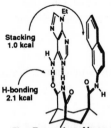

Stacking
1.0 kcal

H-bonding
2.1 kcal

**Fig.1 Energetics of base-pairing
and aryl stacking in CDCl₃**

Xanthene - derived receptors for thymine

We began with placing sizable, polarizable aromatics such as naphthalene on
the framework and measured the response that it provided, presumably by induced
dipoles, to an adenine pinned nearby with hydrogen bonds. In these studies we
made use of organic solvents in which the hydrogen-bonding forces were magnified
for convenient study by NMR titration techniques. These hydrogen-bonding

surfaces could be altered for complementarity to cytosine derivatives[3], to guanosine derivatives[4], and even to thymine derivatives[5] as shown above. For this last case, a xanthene skeleton was devised with the same U-turn as the Kemp triacid, but giving rise to much larger cavities[6].

For the adenine derivatives we were able to show how Watson-Crick vs Hoogsteen base pairing could be controlled by peripheral bulky groups that provided steric effects[7]. These steric effects have counterparts on the purine skeleton as well. For example, any N-substitutents prefer to be directed away from the five-membered ring and expose the adenine skeleton to largely Hoogsteen hydrogen-bonding. (Fig. 2)

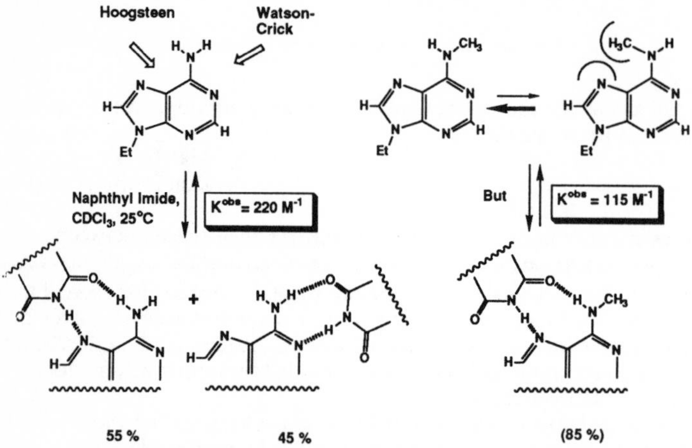

Fig.2 Base-pairing in substituted adenine derivatives

Thus, N-methyl derivatives are at a statistical disadvantage for base pairing to typical imides and intermolecular NOE experiments were able to establish to what extent the numerical disadvantage operated[2].

By placing two imide functions on suitable aromatic spacer we were able to generate molecules with extraordinary affinity for adenines, so much that they provided vehicles for selective adenine transport across simple liquid membranes[8].

match the gentle asymmetry required of adenines chelated by simultaneous Watson-Crick and Hoogsteen base pairing. (Fig. 3)

The carbazole nitrogen provided a chemical handle by which additional recognition elements could be placed onto the skeleton. Specifically, an anionic recognition site, the guanidinium function, developed by de Mendoza[9], Lehn[10], Schmidt[11], could be incorporated. Such molecules showed sufficient affinity for phosphorylated derivatives that stoichiometric amounts of cyclic AMP could be extracted from aqueous solution into organic solvents with these receptors[12]. (Fig. 3) Further elaboration has even permitted the extraction of dinucleotide derivatives[13] and these show great promise as vehicles for delivery of the antisense or ribozyme message across membranes.

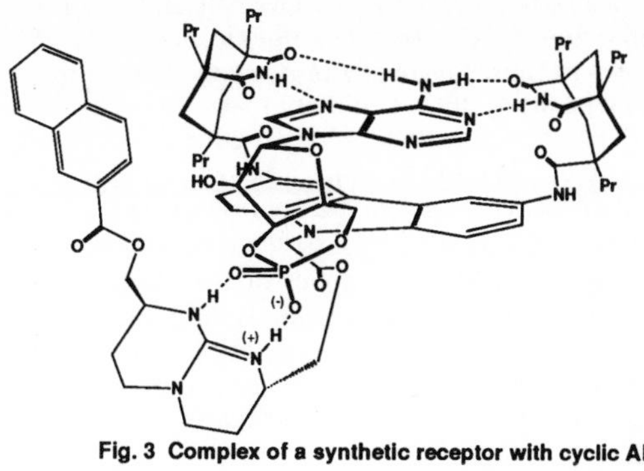

Fig. 3 Complex of a synthetic receptor with cyclic AMP

We were next intrigued by the possibility of covalently attaching adenines to receptors for adenines. At the very least such molecules would have the minimalist requirements of self-complementarity: they could dimerize, they could form higher cyclic oligomers of interest to supramolecular chemistry, and they could even associate in polymeric chains (about which more later). Their most interesting properties dealt with their ability to act as templates for their own formation, that is, to *self-replicate*[14]. This behavior was expressed in the kinetics of the product growth from the coupling of adenosine amines to various activated esters of the imide aromatic receptors. For example, with the biphenyl spacers shown (Fig. 4) sigmoidal product growths were observed which revealed the autocatalytic effect[15]. Coupling of the components gives the J shaped structure which can gather on its exposed surfaces the two reactants from which it is made. This termolecular complex is beautifully poised to reach a relatively effortless transition state for amide formation. This leads to an exact replica of the template which can then dissociate to give the ever increasing amounts of catalysts required for autocatalysis.

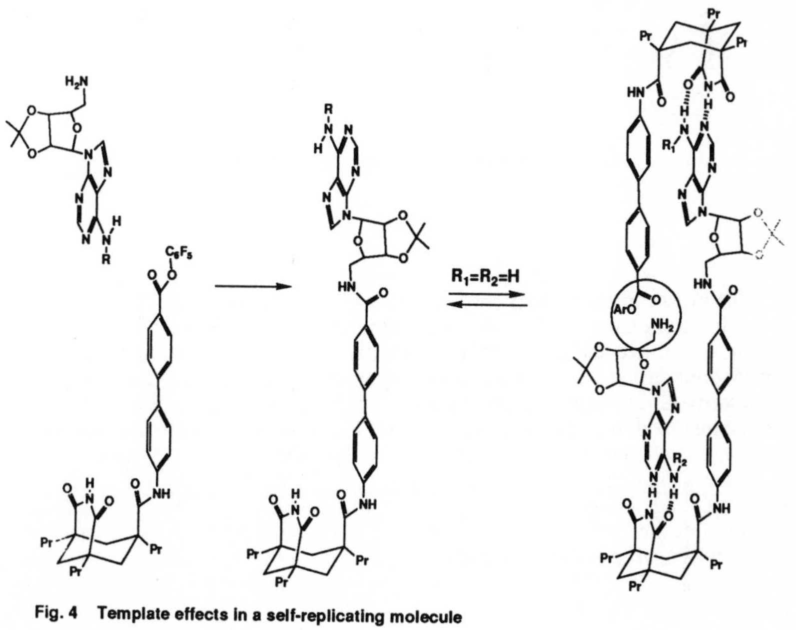

Fig. 4 Template effects in a self-replicating molecule

Such behavior has also been observed by Orgel[16], and von Kiedrowski[17] in simplier nucleic acid derivatives of relevance to prebiotic chemistry. While ours were the first synthetic structures to show this behavior, we encountered the criticism that they could not evolve, that is, the systems could not make mistakes that gave rise to new and more (or less) effective replicators. We overcame this difficulty with N-substituted adenine derivatives. The substituents, a benzyloxycarbonyl and the corresponding ortho nitro derivative, showed the appropriate behavior. Both are replicators and they catalyze not only their own formation, but each others, that is , they make mistakes[18]. (Fig. 5, R, ≠R₂) Moreover such a system shows reciprocity and why not? The essential base pairing information is built into these through the Hoogsteen sense. The molecules can hardly be expected to know whether a nitro group or a hydrogen is in some remote site on their peripheries. They are bound to make mistakes or (as translated into the language of organic chemistry), they lack high selectivity in the self-replicating events.

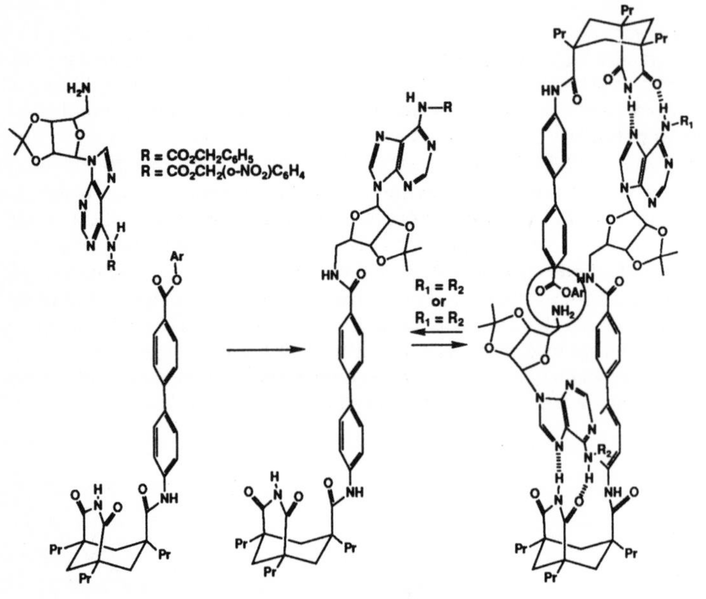

Fig. 5   Competition and reciprocity between replicators

The nitro compound was chosen for its known photolability, that is, irradiation of such structures causes deblocking to the unsubstituted purine nucleus of adenine. The result was simple and predictable. When the carbobenzoxy derivative and the ortho nitro counterpart were competed for a limited amount of active ester, both products formed. The nitro compound was slightly more reactive. When all the ester had been consumed, we irradiated the system. This removed the nitrocarbobenzoxy group from the adenine nucleophile and its corresponding product. Now the addition of more ester gave rise to a new unsubstituted (and uninhibited) replicator. Because of its statistical advantages, its self-replication was enhanced and it rapidly overtook the resources of the system, (the raw materials represented by the active ester). This evolutionary scenario is shown in Fig. 6.

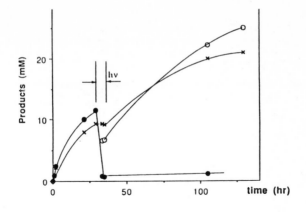

Fig. 6 Irradiation leads to an improved replicator

A different replicating system based on thymine recognition was also devised. Here we made use of the xanthene skeleton because a diaminotriazene was the recognition element. Coupling leads to a self-complementary structure and its shape is insured by the U-turn built into the xanthene dicarboxylic acid module. Again, the product acts as a template to gather the components on its surface for replication[19].

Given the two different replicating systems, we could not resist the urge to combine them. During the coupling reaction all four products were produced, those expected from the previous reactions and the cross-coupling products, hybrids or *recombinants* in which new self-complementary structures were expressed. One of these, the adenine thymine combination is held together by an amide backbone. It begins to resemble modern nucleic acids. Indeed, it is the best replicator we have made to date, and it operates on the same principles involving the template effect in a termolecular complex[20]. The other recombinant, made from two U-turn modules turns out to be "sterile". Even though the structure is self-complementary, it cannot form the cyclic head-to-tail dimer, which is the minimilist structural requirement of these replicators. Rather, it expresses its self-complementarity by forming chains (Figure 7).

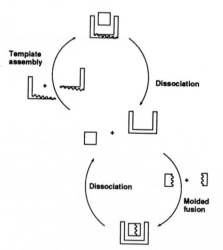

Fig. 7 Crossover leads to fertile and sterile recombinants

Can these results be generalized for other replicators? We believe they can. Consider any recognition event in which two molecules show high affinity for one another (Fig. 8).

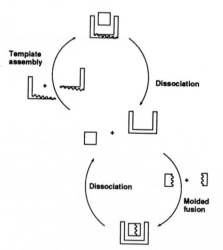

Fig. 8 Recognition and the replication cycle

The convex surface of the cartoon can act as a template for the assembly of the concave suface. Alternatively, the concave surface can act as a mold for fusing the components of the convex structure. This cycle parallels the present replication of nucleic acids one strand acts as a template for the other, which then reciprocates. Recognition is the key to the replication cycle[21].

The cycle can be simplified by a "covalent accident" in which the two complementary molecules become linked at some remote side. If this linkage permits the orientation of the recognition surfaces as shown (Fig. 9), then a minimilist replicator is generated.

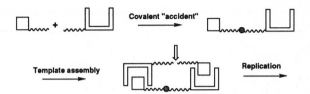

Fig. 9 Minimalist replicators are self-complementary

Note that the actual nature of the replication event is not specified other than shape complementarity. To further explore this notion, we are involved in a collaborative arrangement with F. Diederich in which cyclophane structures that bind aromatic compound in aqueous solution provide the recognition vehicle. By covalently attaching an aromatic as shown in Fig. 10, it should be possible to generate self-replicating systems in water.

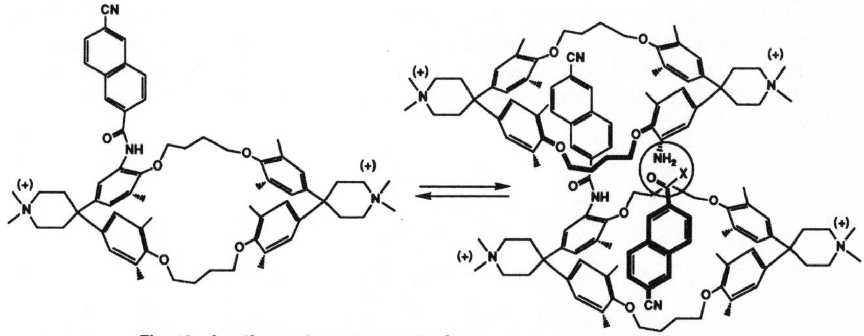

Fig. 10 A self-complementary cyclophane

Elsewhere we have proposed the use of metals and their corresponding ligands for this very purpose[22]. Why not peptides as replicators then? To us it seems utterly reasonable, and we are making progress on this notion in collaboration with A. Rich. Our intent is to couple the glycine-asparagine tetrapeptides on the corresponding octapeptide template (Fig. 11). Again, self-complementarity is the key and here it is represented by the hydrogen-bonded dimers of primary amides.

Ac-Asn-Gly-Asn-Gly-Asn-Gly-Asn-Gly-NH$_2$

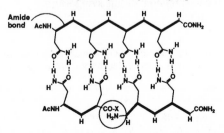

Fig. 11 A self-replicating peptide?

Finally, we consider the remarkable frequency at which self-complementary structures are used in Nature. The subunits of allosteric enzymes and the shells provided by viral coat proteins are popular examples, and the secrets to their self-assembly lie in the *orientation* of the recognition surfaces with respect to one another. One beautiful example is given by guanine tetramers[23] (Fig. 12).

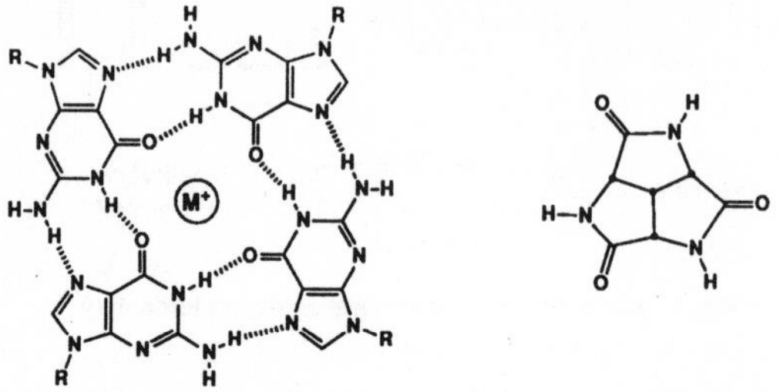

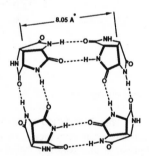

Fig. 12 A G-quartet and a cup-shaped trilactam module

Can such aggregates replicate? Recent experiments with micelles show that indeed they do[24]. The cyclic array of four guanosines around the metal is ultimately due to the readout of hydrogen-bonding information provided by the purine's edges at 90° from one another.

One of the current targets of supramolecular chemistry is the synthesis and assembly of such self-complementary structures[25]. Our own targets involve three dimensional versions and we have identified an unusually promising molecule in that respect. It is the trilactam. This bears the information for appropriate assembly by base pairing into a variety of closed- shell surfaces: tetrahedra, cubes, and even dodecahedra.

Fig. 13 One face of a self-assembling cube

One half of the cube is depicted in Fig. 13; it has the necessary volume to encapsulate sizable structures such as tetramethyladamantane. We will report on its properties in due course, perhaps at the next reunion of this symposium.

ACKNOWLEDGEMENT:

I am grateful to my talented coworkers whose names appear on the original publications. Generous funding was provided by the National Science Foundation.

REFERENCES:

1.   J. Rebek, Jr., *Accounts,* **1984**, *17*, 258-264.
2.   B. Askew, P. Ballester, C. Buhr, K.-S. Jeong, S. Jones, K. Parris, K. Williams and J. Rebek, Jr., *J. Am. Chem. Soc.,* **1989**, *111*, 1082-1090. K.

Williams, B. Askew, P. Ballester, C. Buhr, K.-S. Jeong, S. Jones and J. Rebek, Jr., *J. Am. Chem. Soc.*, **1989**, *111*, 1090-1094.

3. K. S. Jeong and J. Rebek, Jr., *J. Am. Chem. Soc.* **1988**, *110*, 3327-3328.
4. T. K. Park, J.Schroeder and J. Rebek, Jr.,*Tetrahedron*, **1991**, *47*, 2507-2518.
5. T. K. Park, J. Schroeder and J. Rebek, Jr., *J. Am. Chem. Soc.*, **1991**, *113*, 5125-5127.
6. J. S. Nowick, P. Ballester, F. Ebmeyer and J. Rebek, Jr., *J. Am. Chem. Soc.*, **1990**, *112*, 8902-8906.
7. J. Rebek, Jr., K. Williams, K. Parris, P. Ballester and K.-S. Jeong, *Angew. Chem. Int. Ed.*, **1987**, *26*, 1244-1245.
8. T. Benzing, T. Tjivikua, J. Wolfe and J. Rebek, Jr.,*Science*, **1988**, *242*, 266-267.
9. A. Galán, E. Pueyo, A. Salmeron and J. deMendoza, *Tetrahedron Lett.*, **1991**, *32*, 1827-1830.
10. A. Echavarren, A. Galan, J.-M. Lehn and J. deMendoza, *J. Am. Chem. Soc.*, **1989**, *111*, 4994-4995.
11. F. P. Schmidtchen, *Tetrahedron Lett.*, **1989**, *30*, 4493-4496; H. Kurtzmeier and F. P. Schmidtchen, *J. Org. Chem.*, **1990**, *55*, 3749.
12. G. Deslongchamps, A. Galán, J. de Mendoza, and J. Rebek, Jr. *Angew. Chemie. Int. Ed. Engl.*, **1992**, *31*, 61-63.
13. A. Galán, J. de Mendoza, C. Toiron, M. Bruix, G. Deslongchamps and J. Rebek, Jr. *J. Am. Chem. Soc.*, **1991**, *113*, 9424-9425.
14. T. Tjivikua, P. Ballester and J. Rebek, Jr., *J. Am. Chem. Soc.*, **1990,** *112,* 1249-1250.
15. V. Rotello, J. I. Hong and J. Rebek Jr., *J. Am. Chem. Soc.*, **1991**, *113*, 9422-9423.
16. W. S. Zielinski and L. E. Orgel, *Nature*, **1987**, 327-346.
17. G. von Kiedrowski, B. Wlotzka, J. Helbing, J. M. Matzen and S. Jordan, *Angew. Chem. Int. Ed. Engl.*, **1991**, *30*, 423.
18. J. I. Hong, Q. Feng, V. Rotello and J. Rebek, Jr., *Science*, **1992**, *255*, 848-850.
19. T. K. Park, Q. Feng and J. Rebek, Jr., *J. Am. Chem. Soc.*, **1992**, in press.
20. Q.Feng, T. K. Park and J. Rebek, Jr., *Science*, **1992**, in press.
21. J. Rebek, Jr., *Chem. and Ind.*,**1992**, 171-174.
22. M. Famulok, J. S. Nowick, and J. Rebek, Jr. *Acta Chim. Scand.*, in press.
23. R. G. Barr and T. J. Pinnavais, *J. Chem. Phys.*, **1986**, *90*, 328.
24. P. A. Bachman, P. Walde, P. L. Luisi and J. Lang, *J. Am. Chem. Soc.*, **1991**, *113,* 8204.
25. M. Simard, D. Suand, J. D. Wuest, *J. Am. Chem. Soc.*, **1991**, *113*, 4696-4698; Y Ducharme and J. D. Wuest, *J. Org. Chem.*, **1988**, *53*, 5787-5789; M. C. Etter, Z. Urbanczyk-Lipkowska, D. A. Jahn and J. S. Frye, *J. Am. Chem. Soc.*, **1986**, *108*, 5781; For alternative approaches see C. Fouquey, J.-M. Lehn and A. M. Levelut, *Adv. Mater.*, **1990**, *2,* 254; S. J Geib, S. C. Hirst, C. Vincent and A. D. Hamilton, *J. Chem. Soc. Chem. Comm.*, **1991**, 1283; G. M. Whitesides, J. P. Mathias and C. T. Seto, *Science*, in press

Directed Hydrogen Bonding in the Design of New Receptors for Complexation and Catalysis

Robert P. Dixon, Vrej Jubian, Cristina Vicent, Erkang Fan, Fernando Garcia Tellado, Simon C. Hirst and Andrew D. Hamilton

DEPARTMENT OF CHEMISTRY, UNIVERSITY OF PITTSBURGH, PITTSBURGH, PA, 15260, USA

In recent years there has been intense activity in the design of synthetic molecules capable of enzyme-like recognition and binding of small substrates.[1] Two fundamental approaches have been taken. The first has generally involved non-directional binding forces (such as solvophobic, π-stacking and dispersion interactions) in water-soluble cyclophane frameworks.[2] This approach has led to important and quantitative insights into the hydrophobic effect and the enthalpic and entropic contributions of sovent reorganization to binding.[3] However, the weakly orientated nature of the binding interactions has resulted in only moderate substrate selectivity beyond the shape recognition permitted by the cavity. In nature such selectivity is a prerequisite for the chiral recognition and catalytic activity of enzymes and is achieved by hydrogen bonding and electrostatic interactions. The second major approach to artificial receptors makes use of these more directional interactions by incorporating several hydrogen bonding groups into a cleft or cavity of defined geometry.[4] The resulting hosts form strong and selective complexes to those substrates with complementary shape and hydrogen bonding characteristics.[5]

MOLECULAR RECOGNITION BY ARTIFICIAL RECEPTORS

Barbiturate Recognition.

Linking two 2,6-diaminopyridine units through an isophthalate spacer (as in **1**) creates a cavity that is complementary to both the shape and hydrogen bonding features of barbiturates. An X-ray structure of one receptor (**2**) shows a preorganized cavity with all six hydrogen bonding sites directed into the center. Strong complexes to barbiturates ($Ka\sim10^5$ -10^6 M^{-1}) are formed in non-competitive solvents via six hydrogen bonds (**3**). An X-ray structure (**4**) gave details of the distance and orientation of the hydrogen bonds and an analysis of both substrate and receptor analogs has provided an estimate of their strength.[6]

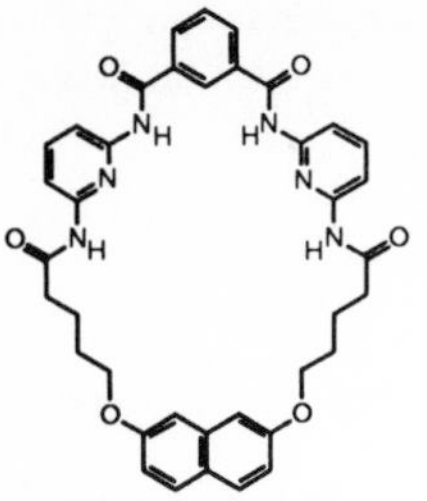

1

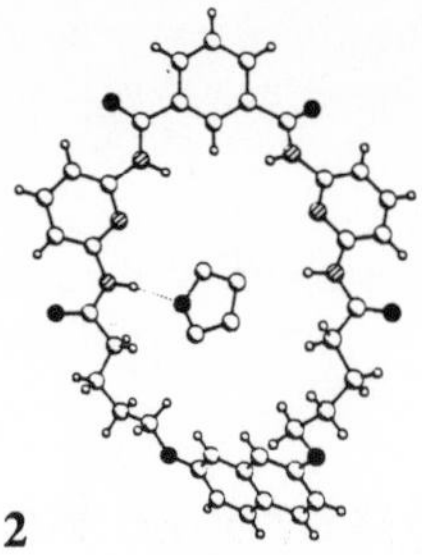

2

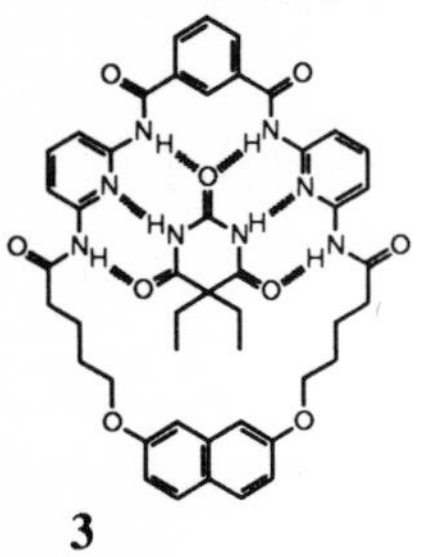

3

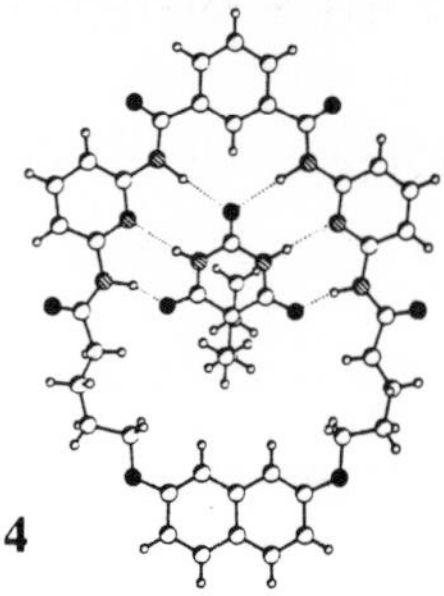

4

Diacid Recognition.

A very simple receptor for diacid substrates was formed from two 2-amino-6-methylpyridine groups separated by a terephthalic acid spacer, **5**. Selectivity is dependent on the spacer length and its fit to the length of the diacid; strongest complexes being formed between **5** and adipic or glutaric acid ($K_a \sim 10^4$ - 10^5 M^{-1}). An X-ray structure of the complex to adipic acid is shown in **6** and confirms the formation of 4 hydrogen bonds between receptor and substrate.[7] This configuration of binding groups is a very simple one that forms the basis of several of our reactivity studies and can be modified for different requirements. For example, a chiral diacid receptor formed from 2,2'-binaphthol-6,6'- dicarboxylic acid forms strong complexes with tartaric acid derivatives **7** with an approximately 3-fold selectivity (K_a=1.0 x 10^6 M^{-1}) for the dipivaloyl D-tartrate compared to the analogous L-tartrate (K_a=3 x 10^5 M^{-1}). The origin of this effect has posed a question that we are currently pursuing since it implies a stabilizing interaction between the binaphthyl π-systems and the pivaloyl-CH$_3$.[8,9]

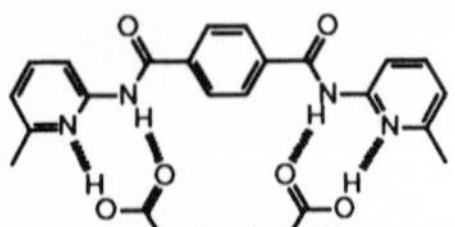

5

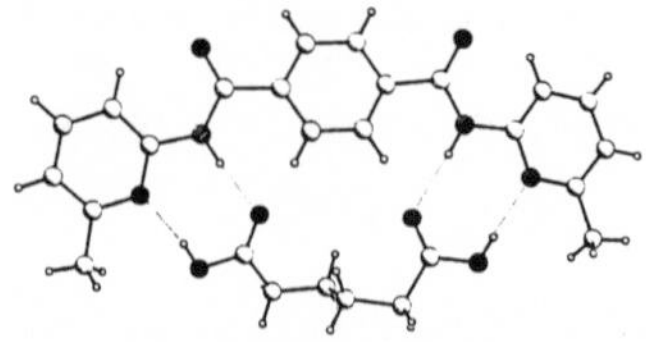

6

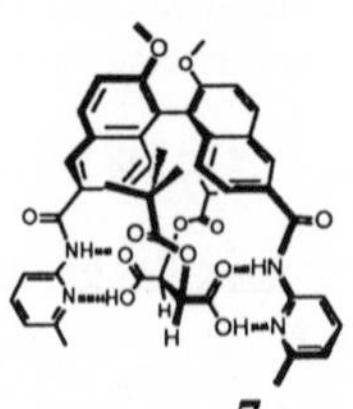

7

Amino Acid Carboxylate Recognition.

In direct analogy to the carboxylate binding pocket of vancomycin[10], we have modified diacid receptors **5**, to permit the recognition of acylamino acid carboxylates. A urea substituent positioned in the 3-position of a benzoate spacer forms two hydrogen bonds to the peptide-CO while a 2-aminopyridine forms a bidentate interaction to the carboxylate terminus, as in **8**. ^{1}H NMR

8   **9**   **10**

experiments established strong binding to N-acetyl-L-proline (K_a=2·6 x 10^3, $-\Delta G_{298}$ = 4·38 kcal mol^{-1}, $-\Delta H_{298}$ = 9·43 kcal mol^{-1} and $-\Delta S$ = 16·9 cal mol^{-1} K^{-1} in H_2O-saturated CDCl$_3$) and participation of the two urea-NH groups.

Phosphate and Carbohydrate Recognition.

        We have recently shown that guanidinium and acylguanidinium derivatives can be used as binding sites for phosphates and carboxylates. This has allowed us to exploit the strong binding available from a charged bidentate interaction as well as extending our recognition studies into more polar solvents. Bis-acylguanidinium receptor **9** is formed in a single synthetic step from the reaction of guanidine and dimethyl isophthalate. Strong complexes (Ka~10^4 -10^5 M^{-1}) involving four hydrogen bonds and additional coulombic interactions, as in **10,** are formed to phosphodiesters in CH$_3$CN.[11] The binding is maintained even in the presence of 10% H_2O suggesting an excellent potential for these receptors in more polar solvent environments. We have also prepared the bis-guanidinium derivative of terephthalic acid **11** and preliminary results show (by analogy to **5** and **6**) that this forms complexes with glutarate and adipate substrates, as in **12**. An X-ray structure of the pimelate salt of **11** (Figure 1) confirms that the acid proton is on the guanidinium and that intramolecular H-bonds from CO to NH are rigidifying the structure.

11   **12**   **13**   **Figure1**

ARTIFICAL RECEPTORS FOR CONTROLLING CHEMICAL REACTIVITY

        Since the original hypothesis of Pauling[12], there have been many attempts to achieve catalysis based on binding site complementarity to a transition state. Most successful has been the development of antibodies elicited against stable molecules that resemble transition state structures.[13,14] The resulting proteins thus contain binding sites that complement the structural, electrostatic and conformational features of the transition state and function as effective catalysts. Catalytic antibodies for a wide range chemical reactions have been developed and accelerations of greater than 10^6 have been seen. The principal limitation of catalytic antibodies is their complexity. At the present time there are no X-ray structures of catalytic antibodies and very little is known about the make up of their active sites or their mechanism of action. Small molecule receptors with complementarity to transition states offer a <u>simple</u> solution to biomimetic catalysis. In addition to our own work, there have been several reports on the rational design of host molecules for transition states.[15,16,17]

Cis-Trans Isomerization of Acylprolines.

The keen current interest in prolyl isomerase enzymes[18] prompted us to consider the chemical microenvironment that would be necessary to influence the rate and equilibrium of amide bond rotation. We focussed on a simple prolyl diacid derivative **14**[19] and showed that the cis-trans equilibrium constant could

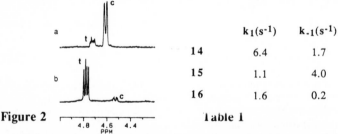

be shifted by more than 30-fold depending on whether the receptor was selective for the cis (as in **15**) or trans (as in **16**) isomer. Figure 2 shows the ^{1}H NMR spectrum of the α-H region for **15**(a) and **16**(b) respectively (assignments being made by intramolecular NOEs from the cis-αH to the fumaramide vinyl-H). These receptors were designed specifically to complex the equilibrium partners. An important question however concerns selective complexation of the transition state in between the cis and trans forms of the amide bond. Stabilization of this twisted amide form would be necessary for catalysis of the isomerization. In preliminary results we have used spin saturation transfer experiments to measure the rate of the forward (k_1) and backwards (k_{-1}) processes. These are collected

	$k_1(s^{-1})$	$k_{-1}(s^{-1})$
14	6.4	1.7
15	1.1	4.0
16	1.6	0.2

Figure 2   Table 1

in Table 1. Significantly the phenyl receptor causes a substantial drop in the rate of the forward and rise in the rate of the backward reactions suggesting a stabilization of the cis and twisted amide forms relative to the unaffected process. Whereas with **16** a large decrease in the rate of the backward reaction indicates a preferential stabilization of the trans isomer.

Intramolecular Diels-Alder Reaction.

Like the acylproline cis-trans isomerization reaction, the intramolecular Diels-Alder (IMDA) involves substantial structural changes in going from starting material to product (Figure 3).[20] The two terminal substituents move from ~6.0Å apart in the required s-cis starting material to 3·4Å apart in the product.[21] Our hope was to discover a simple receptor that would bind to the two substituents (-CO_2H in Figure 3) in an intermediate position and so stabilize the transition state and accelerate the reaction. The simple phenyl receptor (**5**) was complementary to the s-cis ground state, as in **17**, and led to a 10-fold drop in the rate (k_1) of the IMDA (measured by temperature jump methods using ^{1}H NMR in $CDCl_3$). In contrast, the analogous isophthalic-based receptor (which binds to more closely spaced carboxylates as in the putative TS complex, **18**) causes an 30-fold increase in rate of the IMDA (k_1) compared to **17**. The

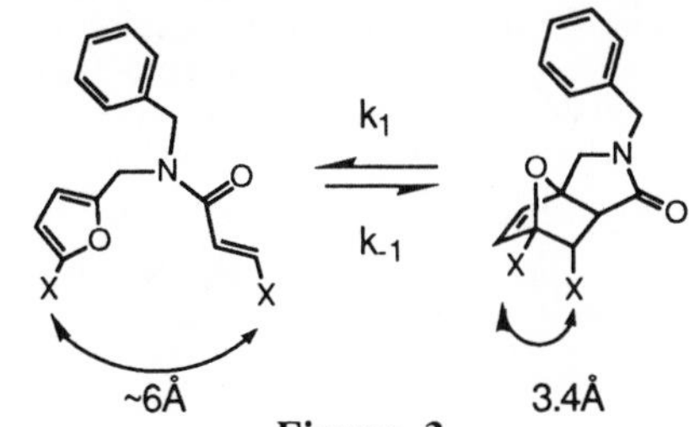

Figure 3

measured rates , collected in Table 2, show little effect of **18** on the retro IMDA (k_{-1}), suggesting that while in **17** only the starting material is stabilized (Figure 4a), in **18** both product and TS are stabilized to similar extents (Figure 4b). This is probably due to the proximity of their carboxylates, compared to the SM, and indicates that more precise design of the receptor is necessary to achieve greater selectivity.[22]

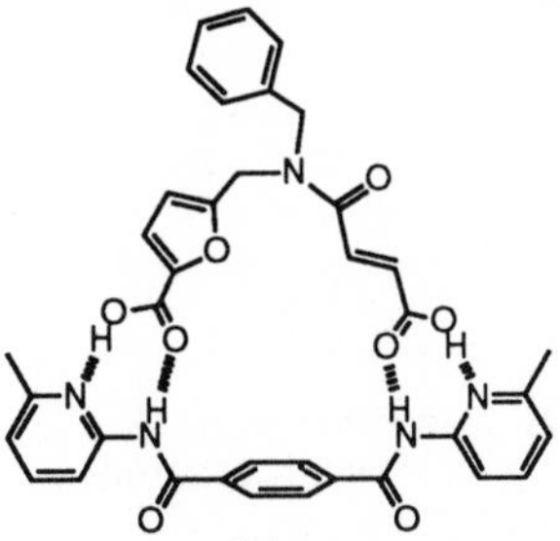

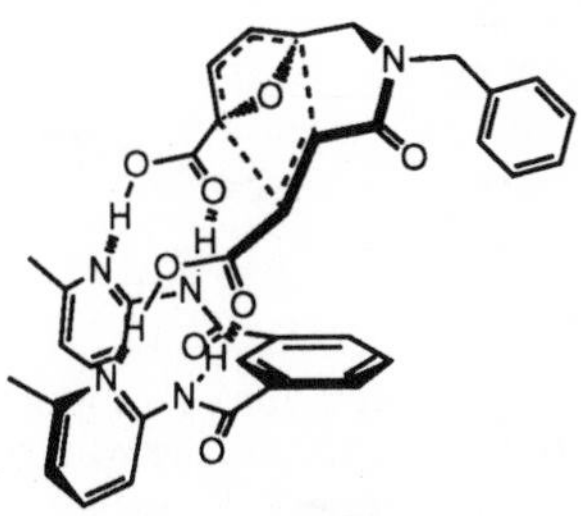

The accelerations seen in both the cis-trans amide bond rotation and the IMDA reaction are modest compared to enzymes or recent catalytic antibodies. However, it should be emphasized that the receptors are extremely simple (one synthetic step!), the results are easily correlated with structural changes and that the origins of the effects appear clear since conformational stabilization is the major influence on the reaction rate with minor contributions from electrostatic or proximity effects.

Receptor	$10^5\ k_1\ (s^{-1})$	$10^5\ k_{-1}\ (s^{-1})$
17	0.48 ± 1.3	1.37 ± 0.32
18	14.0 ± 1.3	2.67 ± 0.52
none	3.96 ± 0.41	4.34 ± 0.50

Table 2. Rates of IMDA reactions   **Figure 4 a   b**

Phosphoryl Transfer Reactions.

Another important goal is the development of catalysts for phosphoryl transfer reactions based on a selective stabilization of the high energy trigonal bipyramidal intermediate. Our first study in this area involved the aminolysis of phosphorodiamidic chloride **19** and exploited the potential complementarity between its trigonal bipyramidal intermediate (TBPI) and the isophthaloyl-bis-aminopyridine receptors, as in **20**. Under pseudo-first order conditions receptor

19   **20**   **21**

22   **Figure 5**

21 caused a 10-fold acceleration in the aminolysis of **19**. This effect was inhibited by addition of a barbiturate which binds strongly to **21** and was absent when simple bis-acylaminopyridines (representing a half of **21**) were used.[23]

The modest acceleration caused by **21** is probably due to the lack of charge neutralization to the monoanion formed in the TBPI. In a second series of receptors we sought to mimic the bis-arginine active site of staphylococcal nuclease by incorporating two guanidinium groups into the receptor. We focussed on the intramolecular phosphoryl transfer reaction of phosphodiester **22**, shown in Figure 5. The key receptor was bis-acylguanidium **9** which binds to monoanionic phosphodiesters as in **10** but was expected to bind more strongly to the bis-anionic TBPI via both hydrogen bonding and charge charge interactions, as in **23**. The pseudo first order rate constants for the phosphoryl

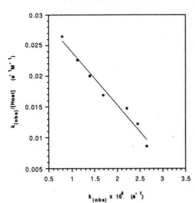

23

Receptor (conc. mM)	k_{obs} x 10^5 sec^{-1}	k_{obs}/k_{uncat}
none	0.038	1
9 (30)	26.5	700
9 (20)	24.5	645
9 (10)	17	450
9 (5)	11	290
Benzoylguanidin. (10)	0.094	2.5

Table 3

Eadie-Hofstee Plot

Figure 6

transfer reaction in CH_3CN/H_2O with lutidine as general base are shown in Table 3. A 700-fold acceleration is seen on addition of a 30mM solution of **9** (as its bis-tetraphenylborate salt) compared to the control reaction in its absence. Simple benzoylguanidinium (**24**) causes only a modest acceleration emphasizing the importance of the bis-cationic binding site. The reaction follows Michaelis-Menton kinetics and shows saturation behaviour as the concentration of the host increases. An Eadie-Hofstee plot (Figure 6) gave a straight line from which values of $k_{cat}=3\cdot8$ x $10^{-4}s^{-1}$ and $K_M =1\cdot2$ x $10^{-2}M$ were calculated.[23]

Acknowledgements

We thank the National Institutes of Health and the Office of Naval Research for financial support of this work.

References

1.   J. M. Lehn, Angew. Chem. Int. Ed . 1988, 27, 88.
      F. N. Diederich, Angew. Chem. Int. Ed .1988, 27, 362.
2.   F. N.Diederich, in Cyclophanes, Royal Society of Chemistry, 1991. K. Odashima, A. Itai, Y.Iitaka and K. Koga, J. Org. Chem. 1985, 50, 4478. C. Kreiger and F. N. Diederich, Chem Ber.1985, 118, 3620.
3.   S. B. Smithrud, T. B. Wyman, and F. Diederich, J. Am. Chem. Soc. 1991, 113, 5420. D. A. Dougherty and D. A. Stauffer, Science (Washington D. C.). 1990, 250, 1558.
4.   A. D. Hamilton, Advances in Supramolecular Chemistry, Vol. 1, G. Gokel, Ed. Jai Press,Greenwich, 1990, p1. J. Rebek, Jr. Acc. Chem. Res. 1990, 23, 399.
5.   B. J. Whitlock and H. W. Whitlock, J. Am. Chem. Soc., 1990, 112, 3910. K. T. Chapman, and W. C. Still, J. Am. Chem. Soc., 1989, 111,

3075. Y. Tanaka, Y. Kato, and Y. Aoyama,. J. Am. Chem. Soc., 1990,
112, 3910. F. Garcia-Tellado, S. Goswami, S. K. Chang, S. Geib and
A. D. Hamilton, J. Am. Chem. Soc. 1990, 112, 7393. T. R. Kelly and
M. P. Maguire, J. Am. Chem. Soc.1987, 109, 6549. J. C. Adrian and C.
S. Wilcox, J. Am. Chem. Soc. 1989, 111, 8055. S. C. Zimmerman and
W. Wu, J. Am. Chem. Soc. 1989, 111, 8054. V. Hegde, J. D.
Madhukar and R. P. Thummel, J. Am. Chem. Soc. 1990, 112, 4549. K.
S. Jeong, T. Tjivikua, A. Muehldorf, G. Deslongchamps, M. Famulok
and J. Rebek, Jr., J. Am. Chem. Soc. 1991, 113, 201. M. Gallent, M. T.
P. Viet and J. D. Wuest, J. Org. Chem. 1991, 56, 2284. T. W. Bell and
J. Liu, J. Am. Chem. Soc. 1988, 110, 3673.

6.   S. K. Chang, D. Van Engen, E. Fan and A. D. Hamilton, J. Am. Chem.
      Soc., 113, 7640 (1991).
7.   F. Garcia-Tellado, S. Goswami, S. K. Chang, S. Geib and A. D.
      Hamilton, J. Am. Chem. Soc., 112, 7393 (1990).
8.   F. Garcia-Tellado, J. Albert and A. D. Hamilton, J. Chem. Soc. Chem.
      Commun., 1761 (1991).
9.   For other examples of attractive aromatic-CH3 interactions see G. D.
      Andreetti, O. Ori, F. Ugozzoli, C. Alfieri, A. Pochini and R. Ungaro, J.
      Inclusion Phenomen. 1988, 6, 523. T. Endo, M. M. Ito, Y. Yamada, H.
      Saito, K. Miyazawa and M. Nishio, J. Chem. Soc. Chem. Commun.,
      1983, 1430.
10.   Kannan, R.; Harris, C.M.; Harris, T. M.; Waltho, J. P.; Skelton, N. N.;
      Williams, D. H. J. Am. Chem. Soc. 1988, 110, 2946.
11.   R. P. Dixon, S. J. Geib and A. D. Hamilton, J. Am. Chem. Soc., 114,
      365 (1992).
12.   L. Pauling, Am. Sci., 1988, 36, 51.
13.   P. G. Schultz, R. A. Lerner, and S. J. Benkovic, Chem. Eng. News,
      1990, 68, 26.
14.   R. A. Lerner, S. J. Benkovic, and P. G. Schultz, Science (Wasington,
      DC), 1991, 252, 659.
15.   J. Wolfe, D. Nemeth, A. Costero, J. Rebek, Jr., J. Am. Chem. Soc.
      1988, 110, 983.
16.   T. R. Kelly, C. Zhao, G. J. Bridger, J. Am. Chem. Soc. 1989, 111,
      3744.
17.   T. Motomura, K. Inoue, K. Kobayashi and Y. Aoyama, Tetrahedron Lett.
      1991, 4757.
18.   R. K. Harrison, and R. L. Stein, Biochemistry, 1990, 55, 4984.
19.   C. Vicent, S. C. Hirst, F. Garcia-Tellado and A. D. Hamilton, J. Am.
      Chem. Soc., 113, 5466 (1991).
20.   M. E. Jung and J. Gervay, J. Am. Chem. Soc. 1989, 111, 5469.
21.   Molecular modelling conducted using the Amber Force field in
      Macromodel v.2.5, W. C. Still, Columbia University.
22.   S. C. Hirst and A. D. Hamilton, J. Am. Chem. Soc., 113, 882 (1991).
23.   P. Tecilla, S. K. Chang and A. D. Hamilton, J. Am. Chem. Soc., 112,
      9586 (1990).
24.   V. Jubian, R. P. Dixon and A. D. Hamilton, J. Am. Chem. Soc., 114,
      1120 (1992).

Structure and Mechanism of Cholesterol Oxidase

Alice Vrielink*, Jiayao Li, Peter Brick and David Blow

THE BLACKETT LABORATORY, IMPERIAL COLLEGE OF SCIENCE, TECHNOLOGY AND MEDICINE, LONDON, SW7 2BZ, UK

*PRESENT ADDRESS: PROTEIN STRUCTURE LABORATORY IMPERIAL CANCER RESEARCH FUND, LONDON, WC2A 3PX, UK

1. INTRODUCTION

The enzyme cholesterol oxidase (3β-hydroxysteroid oxidase, EC 1.1.3.6) catalyzes the oxidation and isomerization of Δ^5-ene-3β-hydroxysteroids with a *trans* A:B ring junction to give the corresponding Δ^4-3-ketosteroids. The oxidation step of the reaction is carried out using FAD as a prosthetic group which is reoxidized by molecular oxygen. The reaction carried out by the enzyme is shown in Figure 1 using cholesterol as the substrate. Although the enzyme exhibits a broad range of steroid specificities, the presence of a 3β-hydroxyl group is essential for activity; 3α-hydroxysteroids are not oxidized by the enzyme.[1-3] The highest enzyme activity is exhibited towards cholesterol, however a high oxidase activity is also observed for steroids without a double bond at C-5 of the steroid ring.

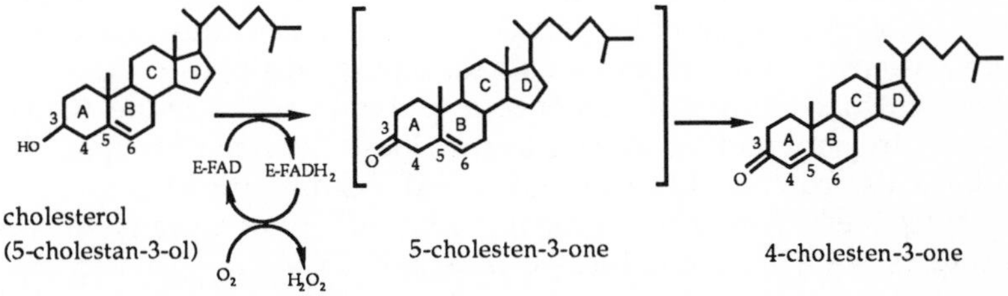

Figure 1. *The enzymatic reaction catalysed by cholesterol oxidase with cholesterol as the steroid substrate. The steroid numbering scheme referred to in the text is shown.*

The enzyme used for this study comes from the soil bacterium *Brevibacterium sterolicum* and is used clinically in the determination of serum cholesterol concentration for the assessment of arteriosclerosis and other lipid disorders. Cholesterol oxidase exists as a monomer of molecular weight 55kDa, (507 amino acid residues) and contains one molecule of tightly bound FAD per molecule of enzyme. The gene sequence of the enzyme has been determined by Ohta and his colleagues.[4]

The X-ray structure of cholesterol oxidase has been determined both without a bound steroid substrate[5] and with the steroid dehydro-isoandrosterone soaked into the native crystals. Both models have been refined to 1.8Å resolution.

The important features of the enzyme needed for specific substrate recognition and catalysis are revealed by studying details of the structures of the enzyme both with and without a bound steroid substrate. These structures enhance the understanding of the interactions made between large hydrophobic substrates and the enzymes with which they interact.

2. DESCRIPTION OF THE OVERALL STRUCTURE

The molecular model of the native enzyme contains 492 amino acid residues and water molecules. The crystallographic R-factor after refinement is 15.3% (on all reflections from 10Å - 1.8Å) with an r.m.s. deviation of bond lengths from ideality of 0.015Å indicating good overall geometry. The overall fold and secondary structure elements are shown in Figure 2. The molecule can be considered to consist of two domains and contains 10 alpha helices and 23 beta strands. The FAD binding domain is made up of three discontinuous regions of sequence: residues 5-44, 226-316 and 462-506 and contains a six stranded beta pleated sheet sandwiched between 3 alpha helices and a 4 stranded beta sheet. This motif has been seen in a number of other FAD binding proteins namely *para*-hydroxybenzoate hydroxylase,[6-8] glutathione reductase[9-11] and lipoamide dehydrogenase.[12,13] A βαβ fold commonly seen in nucleotide binding proteins is involved in binding the adenine ring and interacting with the pyrophosphate oxygen atoms. A common pattern of conserved glycine residues (GXGXXG where X is any residue) is responsible for binding the phosphate oxygen atoms. Further along the sequence a glutamic acid residue (Glu41) makes hydrogen bonding contact to the ribose hydroxyl groups. This residue is also completely conserved in FAD binding proteins.

The steroid binding domain contains two regions of non continuous sequence, residues 45-225 and 317-461. This domain consists primarily of a large six stranded antiparallel beta sheet forming a 'roof' over the active site cavity. A large beta sheet is also present in the structure of *para*-hydroxybenzoate hydroxylase.

The FAD molecule adopts an extended conformation and is involved in 22 hydrogen bond interactions with the protein and surrounding solvent. The adenine and ribose rings are buried in the FAD binding domain. The pyrophosphate oxygen atoms are involved in eight hydrogen bonds, three through main-chain nitrogen atoms of glycine residues which are situated at the amino terminus of helix H1. The negative charge of the oxygen atoms is compensated by the dipole of the helix. The ribityl chain and the flavin ring system extend beyond the FAD binding domain and into what we have defined as the steroid binding domain. One of the two hydrogen bonds formed between the ribityl chain and the protein is between the 2' hydroxyl group and the side-chain nitrogen of Asn119. This residue is conserved in other FAD binding proteins.

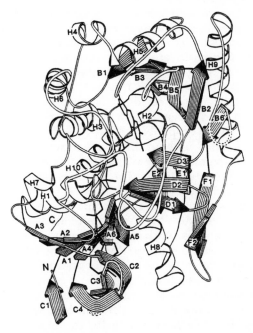

Figure 2. *A representation of the overall fold of cholesterol oxidase showing the strands and helices. The FAD molecule has been included as a stick drawing. The dotted lines indicate loops in the structure which were unable to be interpreted. The strands and helices are labelled as referred to in the text.*

The flavin ring system which has very clear electron density in both the native and the steroid bound structure, shows a number of interesting features. The pyrimidine ring of the flavin group is positioned at the amino terminus of helix H10 with the main chain nitrogen of Phe487 forming a hydrogen bond with O2. This interaction may stabilize the negative charge on the pyrimidine ring after the oxidation step through the helix dipole effect. A helix in the equivalent position with respect to the pyrimidine ring is observed in the structures of *para*-hydroxybenzoate hydroxylase, glutathione reductase and lipoamide dehydrogenase. In the oxidized state it was expected that the flavin ring system would adopt a planar conformation due to the delocalized π electrons. This was however, not observed in the electron density, rather the ring system is twisted with an angle of 17° between the best planes of the benzene ring and the pyrimidine ring.

The active site of the enzyme is located near the flavin ring between the two domains of the molecule. It is surrounded primarily by hydrophobic residues which point inwards towards a large buried cavity. The FAD molecule is positioned at the floor of the cavity which is filled with a well ordered lattice of 13 water molecules forming an ice-like structure. Although the active site is surrounded by many hydrophobic residues the lattice of water molecules filling the cavity makes 13 hydrogen bonding contacts with the protein, three to main chain and ten to side chain atoms. On one side of the cavity, that nearest the benzene ring of the flavin system, are found mainly hydrophobic residues whereas near to the pyrimidine ring a more hydrophilic arrangement of residues is observed. The cavity is approximately 11Å long, large enough to accommodate a steroid ring system[14] of 10Å. The hydrophobic tail on C17 of the steroid backbone cannot be accommodated without displacing a number of hydrophobic loops at the roof of the molecule. We believe this region to provide an entrance to the active site cavity. The residues making up this loop region are found to have higher temperature factors than found in other regions of the structure, suggesting that this region may bé quite mobile. Although a van der Waals surface of these loops shows that the entrance to the cavity is blocked the suggested mobility may allow passage of a substrate into the cavity. Indeed the hydrophobic nature of this channel corresponds to the necessity for an apolar environment for the steroid molecule to pass through.

The active site contains one charged residue, Glu361, protruding into the cavity from strand B4 of the steroid binding domain. In the absence of substrate the electron density for this residue suggests multiple conformations and a high degree of mobility. The occurrence of this charged residue in the active site suggests that it may be involved in the catalytic mechanism Indeed, in the enzyme Δ^5-3-ketosteroid isomerase from *Pseudomonas testosteroni*, an enzyme which carries out the identical isomerisation step as cholesterol oxidase, it has been suggested that the isomerisation involves a 4-6 *cis*-diaxial proton transfer.[15,16] This mechanism would require hydrophobic residues to bind the apolar steroid substrate, an acid residue to polarize the carbonyl group at C3 of the steroid and a basic residue to extract the 4β proton and transfer it to the 6β position.

3. STEROID BOUND STRUCTURE

Although a hypothesis could be formed for steroid binding and catalysis based on the native structure of cholesterol oxidase, a more definitive understanding of the important interactions may be obtained from the structure of a complex of the enzyme with a steroid molecule bound in the active site. A number of factors needed to be considered in order to bind a steroid substrate to the enzyme. Firstly, the observed

K_m for binding cholesterol was far higher than its solubility making it difficult to bind to the enzyme under normal aqueous conditions. Secondly, if a substrate was bound to the enzyme under favourable conditions for catalysis it would be converted to the product. Due to the fact that the reduced enzyme is regenerated by oxygen it is possible to trap the enzyme in the reduced state by eliminating oxygen from the experiment thus disabling it from further catalysis. The steroid dehydroisoandrosterone was found to have enhanced solubility due to the ketone function at C17 of the D ring yet still maintain the features in the A and B rings needed by the enzyme to carry out both the oxidation and the isomerisation steps (see Figure 3). Native crystals of the enzyme soaked in a solution containing dehydroisoandrosterone in the absence of oxygen were observed to change from yellow to colourless after several hours.

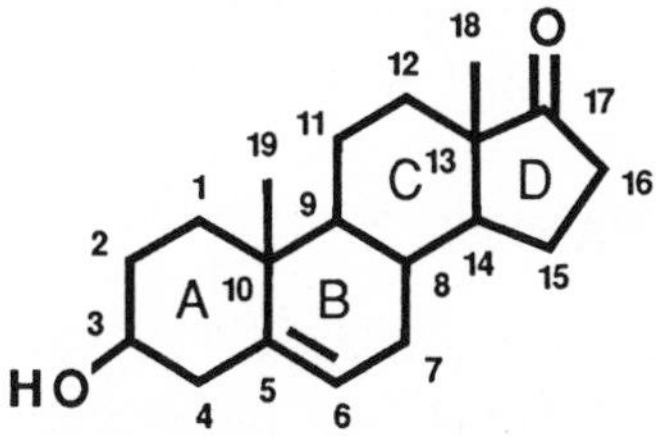

Figure 3. *The chemical structure and numbering scheme of dehydro-isoandrosterone.*

Colourless crystals treated with dehydroisoandrosterone solubilised with alcohol and detergent were then mounted in the absence of oxygen and X-ray data collected to 1.8Å resolution.[17] The difference electron density map showed the steroid bound to the active site of the enzyme having displaced all the water molecules originally observed in the cavity with the exception of one found within hydrogen bonding distance of the disordered Glu361 and the flavin system. The crystallographic *R*-factor after refinement of the complex is 14.5% (on all reflections from 10Å - 1.8Å) and the r.m.s. deviation of bond lengths from ideality is 0.017Å. The orientation of the steroid in the active site cavity is shown in Figure 4. It is most likely that the steroid form bound to the enzyme is that of the substrate and not the product. Although dehydroisoandrosterone could act as a substrate for the enzyme, due to the absence of oxygen each enzyme molecule would only turn over one molecule of substrate; the oxidized form of the enzyme would not be regenerated in the absence of oxygen. Thus the product would leave the active site and a new substrate molecule would enter but not undergo catalysis.

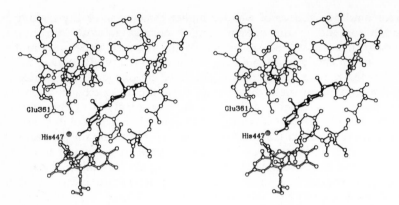

Figure 4. *A stereo view of the steroid substrate, dehydro-isoandrosterone in the active site of the enzyme. The steroid is shown as filled bonds and the flavin ring as open bonds. The water molecule in the active site has been included as a double circle.*

4. POSSIBLE MECHANISMS

Having bound the steroid to the active site it was now possible to observe the interactions made between the protein and the substrate which could be important for catalysis. Although the density for Glu361 is more clearly defined than in the apo structure the average temperature factors for the side chain are still high (30Å^2) suggesting that there is still a large degree of mobility. An intricate hydrogen bonding network is seen around the flavin ring and the 3-hydroxyl group of the steroid molecule (see Figure 5). The conserved water molecule is within hydrogen bonding distance of the hydroxyl group of the steroid as well as to His447 and Glu361 of the protein. The hydroxyl group is also within hydrogen bond distance to O4 and N5 of the flavin ring. The flavin ring system is still observed to be non planar maintaining the hydrogen bond found between O2 of the pyrimidine ring and N-Phe487 in the substrate free structure. These observations suggest a number of possible mechanisms for the oxidation reaction.

Figure 6 shows a mechanism in which the water molecule acts as a base extracting a proton from the hydroxyl group of the steroid which would lose a hydride to N5 of the flavin ring. The resonance forms of the flavin ring would allow the negative charge to be accommodated on O2 of the pyrimidine ring and thus be stabilized by the helix dipole of helix H10. The hydrogen bond observed between the water molecule and His447 suggests that this water molecule acts as a shunt for the proton which, in the absence of other water molecules at the active site, must eventually be accommodated by the histidine. The final reduced form of the cofactor, assuming this mechanism, has gained a proton and two electrons, the second proton being accommodated by His447.

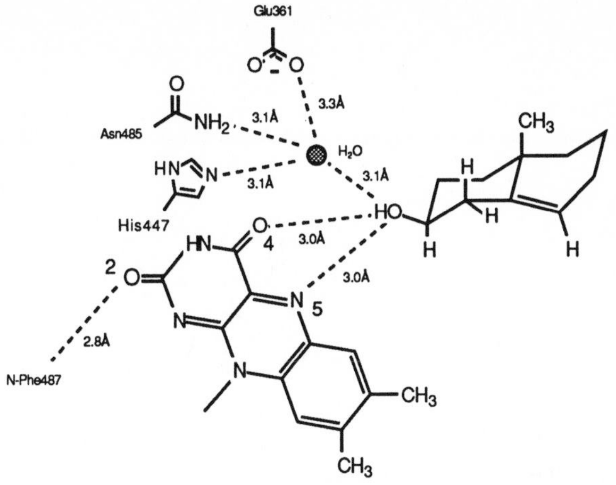

Figure 5. Hydrogen bonding network seen near the hydroxyl group of dehydroisandrosterone in the active site of the enzyme.

A second possible mechanism is shown in Figure 7. Here the oxygen at position 4 of the flavin ring acts as the base, abstracting the 3-hydroxyl proton and, in a concerted manner, a hydride is transferred to the nitrogen at position 5 of the flavin. The final form of the cofactor, by this mechanism, accommodates two protons and two electrons. The residues around the active site are not used in this mechanism but may be thought of as simply forming the scaffold for the steroid to bind near to the flavin ring.

Although the structure of the complex contains all the elements needed to account for the proposed oxidation mechanisms, this is not true for the isomerisation step. It is known that the isomerisation occurs as a separate step from the oxidation since a high oxidase activity is also observed for steroids without a double bond at C5 of the steroid ring.[2,3] As was mentioned above, the enzyme Δ^5-3-ketosteroid isomerase from *Pseudomonas testosteroni* carries out the identical isomerisation step to that of cholesterol oxidase. Nuclear magnetic resonance studies of the Δ^5-3-ketosteroid isomerase-steroid inhibitor complex enabled the location of the bound steroid to be determined and, together with chemical modification studies[18] and affinity labelling studies,[19] identified Asp38 as the proton acceptor and Tyr14 as the proton donor. Site-directed mutagenesis experiments and kinetic analysis further confirmed these residues as important for catalysis.[16]

It has been suggested by Smith and Brooks[20] that the isomerization activity of cholesterol oxidase proceeds via the same mechanism as has been proposed for Δ^5-3-ketosteroid isomerase. The presence of Glu361 in the active site of the enzyme supports this suggestion as this residue might be the base which abstracts the axial hydrogen at C4 of the steroid, according to this mechanism.

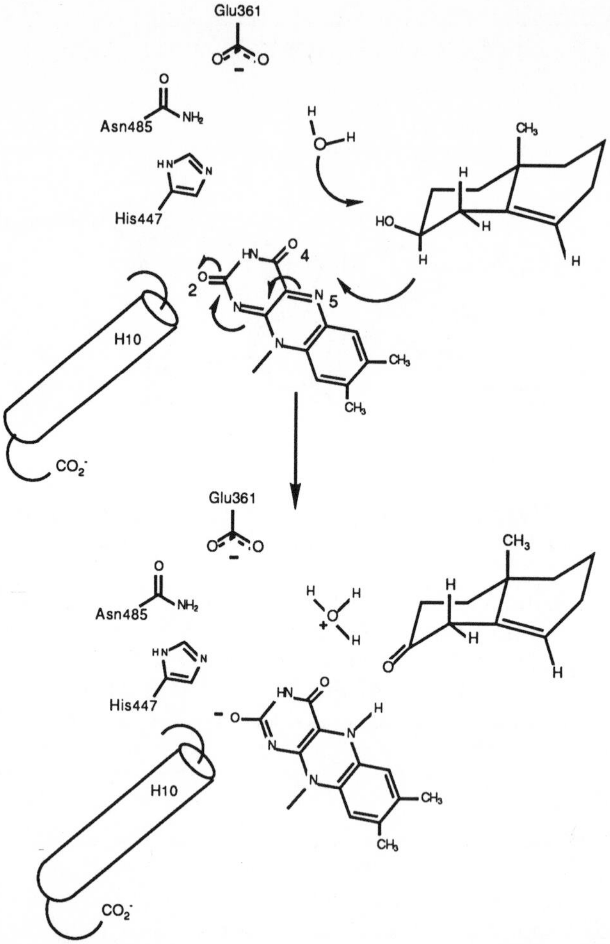

Figure 6. Proposed mechanism for the oxidation by cholesterol oxidase. Only the A and B rings of the steroid and only the flavin moiety of FAD are shown. The conserved water molecule acts as a base extracting the hydroxyl proton. His447 stabilizes the protonated water molecule and may eventually accommodate the proton. The negative charge on the flavin ring is stabilized by the helix dipole effect of helix H10.

In the uncomplexed structure, the side-chain density for Glu361 indicated that it may be mobile. It was expected that the side-chain density of Glu361 would be better defined in the complex structure since the steroid would freeze the side chain in a single conformation. The electron density for the complex, although showing slightly better defined density for the side chain of Glu361 places the carboxylate oxygen atoms over 3.5Å from C4 of the steroid substrate. Modelling studies of the steroid and the glutamic acid side-chain within the observed density indicate however, that a structural change of the steroid during isomerisation might bring them closer together allowing for attack on the C4 axial hydrogen as shown in Figure 8.

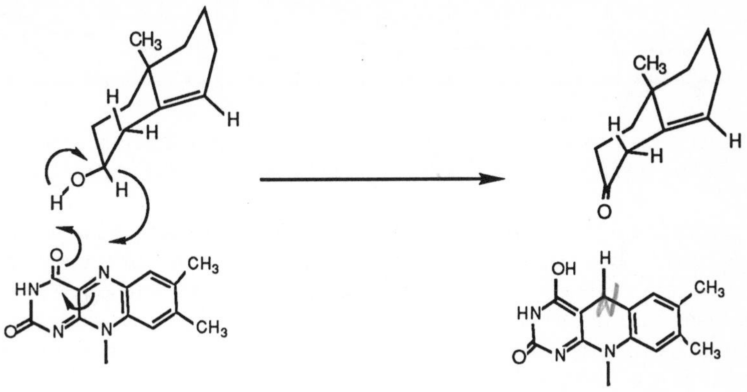

Figure 7. *An alternative mechanism for the oxidation step. The oxygen at position 4 of the flavin ring acts as the base extracting the hydroxyl proton from the steroid substrate. The flavin molecule accommodates two protons and two electrons by this mechanism.*

The complex observed is probably that containing the substrate molecule before oxidation. After the oxidation step the conformation of the A ring will change due to the formation of an sp^2 from an sp^3 hybridized centre at C3. Thus the distance between C4 and the side chain oxygens of Glu361 will differ from that seen in the present model. Further inspection of the steroid bound structure fails to reveal any residue which could act as a proton donor near C6. The glutamic acid may act as both the proton donor and the proton acceptor. For this to occur however the steroid would have to undergo a large movement to place the glutamic acid side chain in close enough contact for attack at C6. These movements and changes in steroid conformation have not yet been observed in the crystal structure. Further experiments to complex transition state analogues and intermediates along the catalytic pathway and study these by crystallography may enable a more detailed mechanistic picture to be established.

The conformation of the steroid nucleus is seriously constrained by the shape of the enzyme's binding pocket. It might be that the energetic consequences of this conformational constraint provide a significant contribution to catalysis by lowering the potential barrier between the two isomeric forms. This matter requires quantitative energetic analysis through detailed molecular modelling.

The structure of the complex reported here, although suggesting a number of possible mechanistic interpretations does not determine unequivocally which the correct one might be. Further structural and biochemical studies must be carried out in order to more fully understand the mechanism.

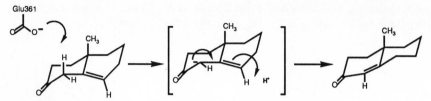

***Figure 8.** Possible mechanism for the isomerisation reaction of cholesterol oxidase using Glu361 as the base for proton extraction.*

A number of site directed mutations and subsequent kinetic analysis may allow these questions to be answered. These mutations include Glu361 and His447. A mutation at Glu361 and subsequent kinetic analyses could test whether this residue is indeed necessary for the isomerisation step. A mutation of His447 would address the questions pertaining to the mechanism of the oxidation step.

ACKNOWLEDGEMENTS

We thank Silvia Onesti for providing many helpful discussions. This project was supported by the Medical Research Council of the United Kingdom.

REFERENCES

1. A.G. Smith & C.J.W. Brooks, Biochem. Soc. Trans., (1975), 3, 675.
2. T. Kamei, Y. Takiguchi, H. Suzuki, M. Matsuzaki and S. Nakamura, Chem. Pharm. Bull., (1978), 26, 2799.
3. Y. Inouye, K. Taguchi, A. Fuji, K. Ishimaru, S. Nakamura and R. Nomi, Chem. Pharm. Bull., (1982), 30, 951.
4. T. Ohta, K. Fujishiro, K. Yamaguchi, Y. Tamura, K. Aisaka, T. Uwajima and M. Hasegawa, Gene, (1991), 103, 93.
5. A. Vrielink, L.F. Lloyd and D.M. Blow, J. Mol. Biol., (1991), 219, 533.
6. R.K. Wierenga, R.J. DeJong, K.H. Kalk, W.G.J. Hol and J. Drenth, J. Mol. Biol., (1979), 131, 55.
7. H.A. Schreuder, J.M. van der Laan, W.G.J. Hol, and J. Drenth, J. Mol. Biol., (1988), 199, 637.
8. H.A. Schreuder, P.A.J. Prick, R.K. Wierenga, G. Vriend, K.S. Wilson, W.G.J. Hol and J. Drenth, J. Mol. Biol., (1989), 208, 679.
9. G.E. Schulz, R.H. Schirmer, W. Sachsenheimer and E.F. Pai, Nature (London), (1978), 273, 120.
10. G.E. Schulz, R.H. Schirmer and E.F. Pai, J. Mol. Biol., (1982), 160, 287.
11. P.A. Karplus and G.E. Schulz, J. Mol. Biol., (1987), 195, 701.
12. A. Takenaka, K. Kizawa, T. Hata, S. Sato, E. Misaka, C. Tamuram and Y. Sasada, J. Biochem., (1988), 103, 463.

13. B. Schierbeek, PhD Thesis, University of Groningen, (1988).
14. B.M. Craven, <u>Acta Crystallogr. sect. B</u>, (1979), <u>35</u>, 1123.
15. F.H. Batzold, A.M. Benson, D.F.Covey, C.H. Robinson and P. Talalay, <u>Advan. Enzyme Regul.</u>, (1976), <u>14</u>, 243.
16. A. Kuliopulos, A.S. Mildvan, D. Shortle and P. Talalay, <u>Biochemistry</u>, (1989), <u>28</u>, 149.
17. J. Li, P. Brick, A. Vrielink and D.M. Blow, manuscript in preparation.
18. J.R. Ogez, W.F. Tivol and W.F. Benisek, <u>J. Biol. Chem.</u>, (1977), <u>252</u>, 6151.
19. R.M. Pollack, S. Bantia, P.L. Bounds and B.M Koffman, <u>Biochemistry</u>, (1986), <u>25</u>, 1905.
20. A.G. Smith and C.J.W. Brooks, <u>Biochem. J.</u>, (1977), <u>167</u>, 121.

Catalytic Mechanism and Active-site Structure of Methylaspartate Ammonia-lyase: Possible Involvement of an Electrophilic Dehydroalanine Reaction Centre

Nigel P. Botting[1], Mahmoud Akhtar[1], Catherine H. Archer[1], Mark A. Cohen[2], Neil R. Thomas[1], Sayed Goda[3], Nigel P. Minton[3] and David Gani[1]

[1] DEPARTMENT OF CHEMISTRY, THE PURDIE BUILDING, THE UNIVERSITY OF ST. ANDREWS, ST. ANDREWS, FIFE, KY16 9ST, UK
[2] DEPARTMENT OF CHEMISTRY, THE UNIVERSITY, SOUTHAMPTON, SO9 5NH, UK
[3] MOLECULAR GENETICS GROUP, DIVISION OF BIOTECHNOLOGY, PHLS CENTRE FOR APPLIED MICROBIOLOGY, PORTON DOWN, SALISBURY, WILTSHIRE, SP4 0JG, UK

1. Introduction

3-Methylaspartate ammonia-lyase (β-methylaspartase, EC 4.3.1.2) catalyses the reversible α,β-elimination of ammonia from L-*threo*-(2S,3S)-3-methylaspartic acid (1) to give mesaconic acid (Scheme I).[1]

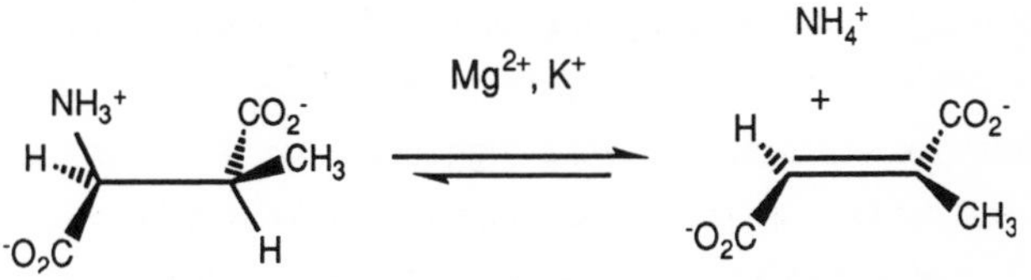

Scheme 1. The Methylaspartase Reaction

The enzyme follows glutamate mutase in the catabolism of glutamate in *Clostridium tetanomorphum*[1] and a number of other species.[2,3] The enzyme from *C. tetanomorphum*, which is the best studied, was shown to deaminate (2S)-aspartic acid and a number of 3-alkyl homologues,[4] and also, the L-*erythro*- (2S,3R)-diastereomer of methylaspartic acid.[1] The enzyme was reported to possess a tetrameric (AB)$_2$ structure, M$_r$ 100,000[5,6] and was demonstrated to require monovalent as well as divalent cations for activity.

In early work on the mechanism of deamination, it was noted that the Clostridial enzyme was able to catalyse the exchange of the C-3 hydrogen atom of the physiological substrate with hydrogen derived from the solvent, at greater rates than the overall deamination reaction, under some conditions.[7,8] No primary deuterium isotope effect was detected for the deamination reaction over the pH range 5.5-10.5 and on the basis of these observations, together with the reported ability of the enzyme to process both diastereomers of L-3-methylaspartic acid, a mechanism involving the intermediacy of a carbanion was proposed [see Hansen and Havir[9] for a review of the early work]. Since Bright's work, the methylaspartase system has been regarded as the archetypal example of an enzyme which operates *via* a carbanion elimination mechanism, and indeed, studies of the related systems, L-aspartase,[10] phenylalanine ammonia-lyase[11] and argininosuccinate lyase[12] have revealed the operation of similar elimination mechanisms.

Research in our own laboratories led us to investigate the kinetics of the methylaspartase system using a range of substrate analogues[13,14] and, furthermore, prompted an examination of the primary deuterium isotope effects for the deamination of these substrate analogues.[15] In contrast to the reported findings,[7] we found that (2S,3S)-3-methylaspartic acid, the physiological substrate, showed a significant isotope effect of ~1.7 on V and V/K over a wide pH range. It was also shown that the C-3 hydrogen exchange reaction displayed a significant primary deuterium isotope effect.[16] Together with the results of product inhibition experiments and ^{2}H/^{1}H-^{15}N/^{14}N-double isotope fractionation experiments[17] it was suggested that the chemical elimination process was concerted and was followed by a slow step in which ammonia or ammonium ion was relocated within the active site of the enzyme prior to the release of the products.[16]

Here we describe our recent efforts to identify the slow post-deamination step, to further unravel the catalytic mechanism of the deamination process and to gain structural information on the active-site of the enzyme.

2. Cloning, Sequencing and Expressing β–Methyl-aspartase.

In order to facilitate a number of experiments designed to identify and characterise the active-site of the enzyme, a programme to clone and express the β-methylaspartase gene from *Clostridium tetanomorphum* in *E. coli* was initiated. Since it had been reported that the enzyme possessed an A_2B_2 tetrameric structure, the first objective was to resolve and purify each of the protomers for N-terminal amino acid analysis. In the event we were unable to confirm the tetrameric structure of the protein and in our hands the enzyme behaved as an A_2 homodimer (M_r 100,000, G3000 gel exclusion FPLC) with subunits of 49,000 Daltons, as determined by SDS-polyacrylamide electrophoresis.

N-Terminal Edman degradation of a highly purified sample of methylaspartase allowed the identity of the first 26 amino acids in the protein to be determined. There were no complications in the sequence analysis that might indicate that more than one polypeptide was present, a result which is in complete accord with an A_2 homodimeric protein structure. The sequence of the N-terminal region was MKFVDVLNTPGLTGFYFDDQAAIKKG.

Using the protein sequence, a 77-mer oligonucleotide probe was prepared for *Clostridial* DNA hybridisation experiments. This oligonucleotide, (5'-ATGAAATTTGTTGATGTTTTAAATACTCCAGGATTAACTGGATTTTATTTTGA TGATCAAGCTGCTATAAAAAAAGG-3', corresponded to a sense DNA strand capable of encoding the derived N-terminal peptide sequence and incorporated, in positions of codon degeneracy, those nucleotide bases most frequently utilised in clostridial genes.[18] The 5'-end ^{32}P-labelled probe hybridised strongly to *Clostridial* DNA and thus the identification of the methylaspartase gene was facilitated.

A 2.2 Kb *Clostridial* DNA *Hind*III restriction fragment was shown to contain the complete β-methylaspartase gene. Accordingly, the sequence of the *Hind*III fragment (Figure 1) was determined in its entirety on both DNA strands. Translation of the nucleotide sequence indicated that two open reading frames (ORFs) were present. The first, ORF A, was apparently incomplete, initiating external to the *Hind*III fragment and terminating with a UAA codon at position 627. The second, ORF B, corresponded to that previously identified as encoding β-methylaspartase. ORF B started with an AUG codon at position 756, preceded by a sequence bearing a strong resemblance to the ribosome binding sites of clostridial genes,[18] and terminated with a UAA translational stop codon at position 1995. The deduced primary sequence of the ORF B polypeptide (Figure 1) contains 413 amino acid residues, $M_r = 45, 539$.

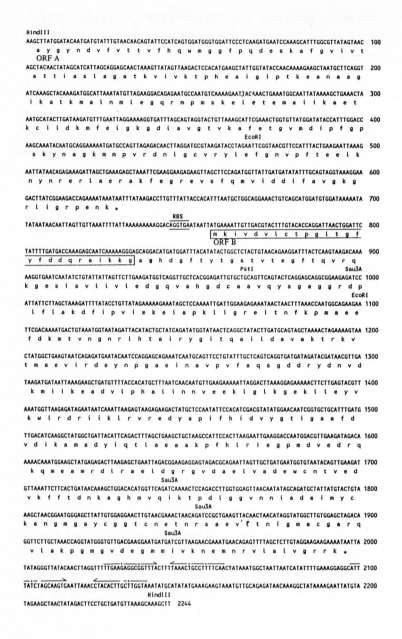

Figure 1. Sequence of the *Hind*III fragment

The *Hind*III fragment was subcloned into both pMTL21 and pMTL32, in the former case such that transcriptional readthrough of the gene could occur from the vector derived *lac* promoter. *E. coli* cells carrying the two resultant plasmids (pSG4 and pSG5, respectively) were cultivated in YT media and the levels of recombinant enzyme produced were assessed in sonic extracts of harvested cells using the standard assay. pSG4, which yielded methylaspartase at a level of ~6% of total soluble protein, was used to prepare recombinant methylaspartase for characterisation and comparison with the authentic enzyme.

3. Purification and Characterisation of Recombinant β-Methylaspartase from *E. coli.*.

pSG4 was cultivated in 7 litres of YT medium to give 95 g of wet cell paste. The enzyme was purified from 10g of frozen cell paste in 36% overall yield to give 4 mg of homogeneous protein (as judged by SDS-PAGE) using protocols similar to those used for purifying the enzyme from *C. tetanomorphum,* see Table I. The recombinant protein was indistinguishable from the authentic protein purified from *C. tetanomorphum* on the basis of SDS-PAGE analysis, Figure 2 and on the basis of its kinetic properties, see below.

Size and Subunit structure. Translation of the determined nucleotide sequence shows that methylaspartate ammonia-lyase is composed of 413 amino acid residues. The predicted amino acid composition (Table II) exhibits a high degree of agreement with that determined experimentally on protein hydrosylates by Williams and Libano[19] and by Hsiang and Bright.[5] The predicted molecular weight of the protein (M_r 45, 539) is consistent with that estimated by SDS PAGE (M_r 49, 000) for enzyme purified from *C. tetanomorphum*, Figure 2. The established molecular weight of the protein, M_r ~100,000 (Hsiang & Bright, 1967) indicates that the active enzyme is a simple A_2 homodimer rather than an A_2B_2 tetramer as was originally thought.[5,6]

Table I: Purification of Recombinant Methylaspartase.

Fraction	Volume (ml)	Protein (mg)	Activity Total (units)	Activity Specific (units mg^{-1})	Purification Factor	Yield (%)
Crude Extract[a]	24.0	201.0	856	4.3	1.0	100
Resuspended Acetone Pellet	2.8	80.3	1256	15.6	3.6	147
Reduced G-150 Fraction	3.0	30.3	1004	33.1	7.7	117
Anion Exchange FPLC Fraction-1	3.3	10.2	480	47.1	11.0	56
Anion Exchange FPLC Fraction-2	2.9	4.0	310	77.5	18.0	36

[a]Yields are calculated from the activity present in the crude extract.

Assays on this fraction underestimate the actual activity, due to the presence of a methylfumarase activity.

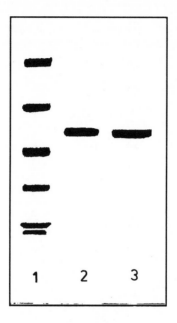

Figure 2. SDS-polyacrylamide gel electrophoresis of the natural and recombinant proteins. Lane 1, rabbit muscle phosphorylase b (M_r 97,400), BSA (M_r 66,200), hen egg white ovalbumin (M_r 45,000), bovine carbonic anhydrase (M_r 31,000), soyabean trypsin (M_r 21,500) and hen egg white lysozyme (M_r 14,400); lane 2, recombinant methylaspartase; lane 3, methylaspartase isolated from *C. tetanomorphum*.

Kinetic properties.

The recombinant protein showed very similar kinetic properties to the protein isolated from *C. tetanomorphum*, Table III. The values of K_m for each protein with (2S,3S)-3-methylaspartic acid as the substrate were identical although the recombinant protein showed a slightly lower specific activity.

The primary deuterium isotope effects for V and V/K for the two enzymes were also the same. The effects of the concentration of the monovalent cationic cofactor, K^+, on the kinetic constants and on magnitude of the deuterium

Table II. Comparison Of the Determined and Deduced Amino Acid Composition of β-Methylaspartase.

Amino Acid	Deduced[a]	Determined[a]		Deduced[b]	Determined[b]
Alanine	41	43	43	4	5
Arginine	23	23	29	2	2
Asparagine	18	-	-	3	-
Aspartic Acid	33	-	-	5	-
(Asp + Asn)	51	54	54	8	8
Cysteine	7	8	8	0	0
Glutamine	12	-	-	1	-
Glutamic Acid	29	-	-	3	-
(Glu + Gln)	41	45	44	4	4
Glycine	38	42	40	2	8
Histidine	7	8	9	0	0
Isoleucine	29	30	29	2	2
Leucine	27	30	29	0	1
Lysine	28	33	36	2	2
Methionine	17	16	18	1	1
Phenylalanine	14	16	14	1	1
Proline	13	12	13	2	2
Serine	6	9	7	1	1
Threonine	22	22	22	1	1
Tryptophan	2	2	-	0	-
Tyrosine	12	-	-	2	1
Valine	35	37	36	5	3

[a]For entire protein.[5]   [b]For an N-ethylmaleimide peptide,[6] see text for details.

isotope effects were identical for both enzymes. The K^+ concentration dependence of these parameters is a very sensitive test for the binding and the debinding orders of the substrate, products and metal ions.[16]

Interestingly, the recombinant protein did catalyse the slow conversion of mesaconic acid to (2S,3R)-3-methylaspartic acid in the presence of ammonia, as well as the rapid conversion of mesaconic acid to the natural substrate, (2S,3S)-3-methylaspartic acid. The formation of the (2S,3R)-epimer, a formal *syn*-addition process is catalysed by the enzyme purified from *C. tetanomorphum*. The finding that the recombinant protein displays an identical activity rules out the possibility that the operation of a contaminating protein in *C. tetanomorphum* was responsible for the formation (or deamination) of the minor (2S,3R)-epimer in the work originally reported by Barker and coworkers.[1] Note that the original kinetic parameters quoted by Barker for the (2S,3R)-epimer differ substantially from findings in our laboratory (with the native enzyme) using very pure synthetic (2S,3R)-3-methylaspartic acid (C. H. Archer, N. R. Thomas and D. Gani, unpublished results). We find that K_m is 30 mM and $V_{Threo}/V_{Erythro}$ is 20. The values quoted by Barker were 0.6 mM and 100 respectively.[1]

Location of an active site peptide. Examination of the primary structure of methylaspartase (Figure 1) indicates there are seven cysteine residues, Table II. This number is consistent with the fourteen cysteine residues per mole of enzyme that could be titrated by photooxidation as reported by Williams and Libano.[19] However, Wu and Williams also showed that eight uniformly labelled peptides were formed upon the treatment of methylaspartase with [^{14}C]-N-ethylmaleimide followed by tryptic digestion, and assumed that eight cysteine residues in the protein had been labelled.[6] This result led these researchers to believe that their earlier estimate of the number of Cys residues in the molecule had been too low. Bright also provided evidence for the presence of sixteen Cys residues per mole of enzyme on the basis of p-chloromercuribenzoate titrations.[5] Further work in Williams' laboratory allowed an active site peptide to be isolated.[6] When the enzyme was treated with unlabelled N-ethylmaleimide in the presence of substrate, dialysis of the protein solution gave an active enzyme. Treatment of the dialysed enzyme with [^{14}C]-N-ethylmaleimide caused inactivation and treatment of this labelled protein with trypsin gave a

Table   III: Variation of Kinetic Parameters for (2S,3S)-3-Methylaspartic acid with K$^+$ Concentration[a] for Natural (C) and Recombinant (R) Enzyme.

| [KCl]/mM | Enzyme | (2S,3S)-3-Methylaspartic Acid | | (2S,3S)-[3-^{2}H]-3-Methylaspartic Acid | | | |
		K_M/mM	V_{max}[b]/10^{-6} mol dm^{-3} min^{-1}	K_M/mM	V_{max}[b]/10^{-6} mol dm^{-3} min^{-1}	D_V	$D_{(V/K)}$
0.3	C	2.80±0.8	309	2.63±0.5	274	1.13	1.06
0.3	R	2.54±0.5	231	2.38±0.4	195	1.19	1.11
1.6	C	2.37±0.20	654	2.35±0.25	385	1.70	1.68
1.6	R	2.17±0.15	462	2.14±0.20	283	1.63	1.70
50.0	C	0.67±0.07	2089	0.67±0.07	2089	1.00[c]	1.00[c]
50.0	R	0.61±0.09	1597	0.61±0.09	1597	1.00[c]	1.00[c]

[a]Incubations contained 0.5 M Tris (pH 9.0), 0.02 M MgCl$_2$, substrate and KCl, in a total volume of 3 ml. Reactions were run at 30±0.1°C.

[b]Corrected for one unit of natural enzyme assayed at pH 9.0. This batch of recombinant enzyme showed 70-75% of the specific activity of the natural enzyme.

Errors on V_{max} values are <8%.

[c]Both isotopomers lie on the same on plot.

single active site labelled peptide. This peptide corresponded to one of the eight uniformly labelled peptides obtained from the earlier experiment and was assumed to contain cysteine. The amino acid composition of the active site peptide was determined by total acid catalysed hydrolysis and the peptide was estimated to be a 39-42-mer.[6] Note that the structure of the modified cysteine hydrolysis product, S-succinocysteine was not proved in this work.

When we compared the possible tryptic peptide sequences and amino acid compositions for 35-45-mers for each of the seven cysteine residues in methylaspartase with the reported amino acid composition of the active site N-ethylmaleimide labelled peptide, there was no correlation whatsoever.
This analysis indicates that the active site peptide does not contain a cysteine residue and that the other seven uniformly labelled peptides detected in Wu and Williams experiment were (probably) cysteine containing peptides.

When all of the possible tryptic peptide sequences and amino acid compositions for 35-45-mers were compared with the reported amino acid composition of the active site N-methylmaleimide labelled peptide, relaxing the requirement for the presence of cysteine, there was one very good correlation. This possible peptide, a 37-mer, which spans Lys-147 to Lys-183 (Figure 1) contains a unique serine residue, Table II. Note that this peptide would arise from cleavage at Arg-146 and Lys-183. There is another site for trypsin cleavage at Lys-144 and cleavage here would give a 39-mer containing an extra Thr and Arg residue, see Figure 1 and Table II.

The correlation of the deduced 37-mer peptide composition with the amino acid composition of the tryptic peptide was perfect for Arg, Asp and Asn, Glu and Gln, His, Ile, Lys, Met, Phe, Pro, Ser, Thr, and very good for Ala. Indeed, glycine was the only amino acid which correlated very poorly. On the basis of this analysis it seems almost certain that the so-called active site cysteine containing peptide isolated by Wu and Williams actually contained serine or a post-translationally modified serine residue.

4.   Catalytic Mechanism
To understand why the product of the serine codon should react with N-

ethylmaleimide is not easy if serine is the final product. The β-hydroxyl group of serine does not usually react with Michael acceptors to give conjugate addition products. Also, the finding by Bright that treatment of the enzyme with eight molar equivalents of p-chloromercuribenzoate caused peptide chain cleavage is curious.[5]

Recent studies in our laboratory have shown that phenylhydrazines are potent irreversible inactivators of methylaspartase although negatively charged nucleophiles such are borohydride and cyanide are not so effective. Substrate protects against inactivation by phenylhydrazines (Akhtar & Gani, unpublished results) and the findings suggest that the enzyme may contain an electrophilic residue, such as a dehydroalanine residue. Interestingly, a dehydroalanine residue has been implicated in the deamination of phenylalanine and histidine by phenylalanine[20] and histidine[21] ammonia-lyases.

Further substantial support for the operation of a dehydroalanine residue which can act as covalent binding site for ammonia is provided by the results of our previous kinetic studies.[15-17] For example, the ^{15}N-isotope effect for V/K for the deamination of (2S,3S)-3-methylaspartic acid at pH 6.5 increased from 1.0255 to 1.0417 when deuterium was introduced at C-3 of the substrate.[17] This result indicates that a concerted deamination mechanism operates.[22] On the other hand, ammonia was shown to increase the rate of C-3 substrate hydrogen exchange with the solvent, without itself becoming incorporated into the substrate pool,[7,16] a result most easily accommodated by a stepwise carbanionic mechanism. Nevertheless, the exchange reaction displayed a similar primary deuterium isotope effect to that for deamination reaction over a wide pH range (6.5-9.0), over which the ratio of deamination to exchange was known to vary by a factor of 5.[16] These results led us to propose that there was an extra binding site for ammonia, in addition to the product binding site, and that under certain conditions the vacation of this extra binding site was kinetically significant. Independent support for this notion was obtained from the demonstration that ammonia acted as a non-linear *uncompetitive* product inhibitor for the deamination process.[16] This result indicates that; A, ammonia or ammonium ion binds to the enzyme to introduce an irreversible step (hence, uncompetitive inhibition, a very rare and informative action for a product), and that; B, more than one molecule of the species can bind to the same form of the enzyme. The most simple mechanism consistent with the data is depicted in Scheme II.

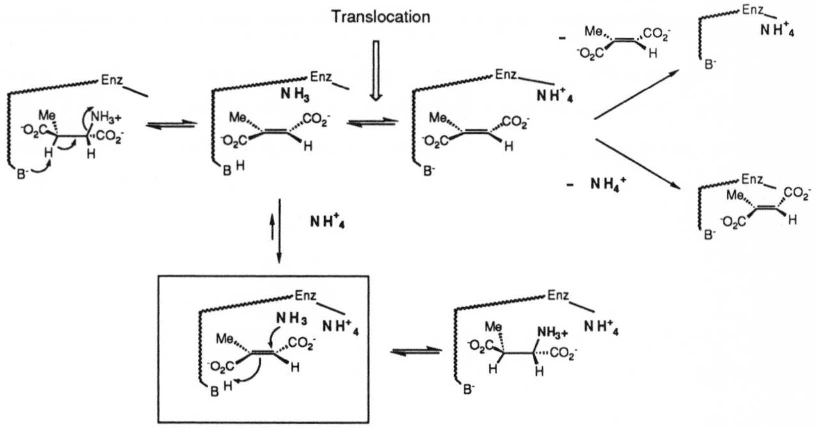

Scheme II. Presence of two binding sites for ammonia. Formation of the boxed intermediate causes uncompetitive inhibition.

If Ser-173 is indeed converted to a dehydroalanine residue, the post-translational modification in the recombinant protein must be catalysed by the enzyme itself. This must be true (unless the host possesses the necessary processing activities) because only one complete ORF is present in the cloned clostridial fragment. The mechanism for catalysis would then follow on from an elaboration of Scheme II where one of the ammonia binding sites is the dehydroalanine residue, as depicted in Scheme III.

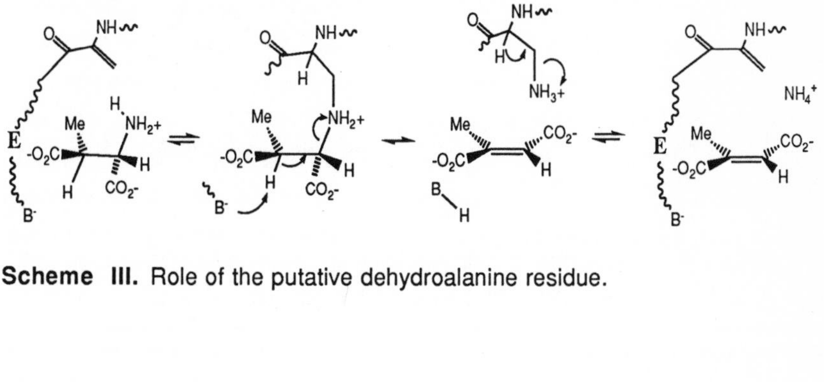

Scheme III. Role of the putative dehydroalanine residue.

A dehydroalanine residue would be expected to react with hydrazines, to give inactive enzyme, in the same manner as substrate and ammonia, *via* β-addition of the nitrogen nucleophile to the α,β-unsaturated amide.

The dehydroalanine residue is also an acylated enamine and, as such, would be nucleophilic at the β-carbon atom and prone to reactions with electrophiles. To this end, the reaction of the dehydroalanine residue with mercury salts would be expected to result in amide bond cleavage as indicated in Scheme IV.

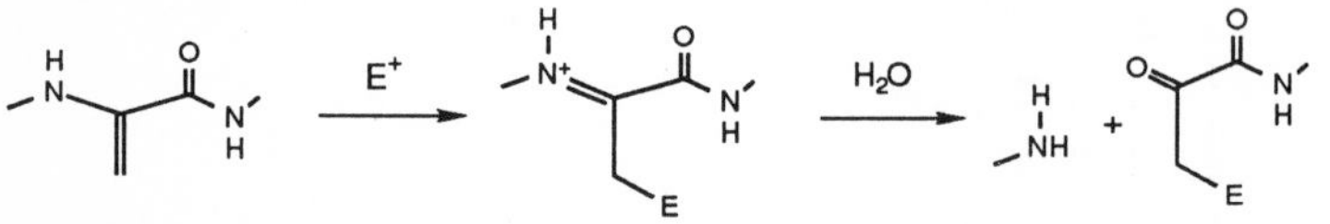

Scheme IV. Electrophile activated peptide chain cleavage.

The product of ammonia addition to the putative dehydroalanine residue, a 2,3-diaminopropionic acid residue, whether formed directly or through its elimination from an intermediate substrate complex, would be expected to add to N-ethylmaleimide *via* its β-amino group. The reaction is almost identical to the microscopic reverse of the catalytic elimination reaction in which the maleimide replaces a fumarate, Scheme V.

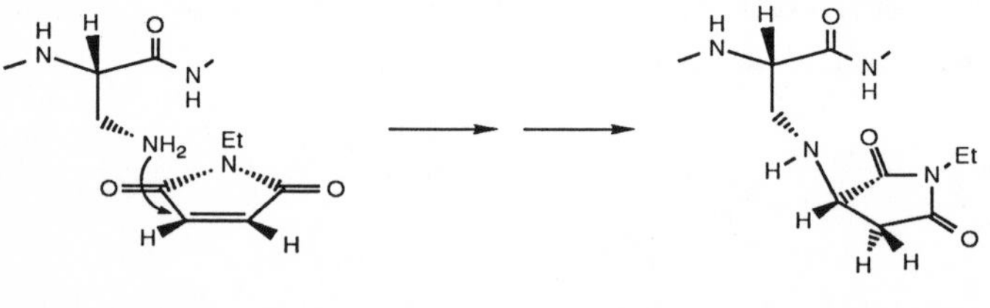

Scheme V. Reaction of a catalytic intermediate, the 2,3-diaminopropionic acid residue, with N-ethylmaleimide.

Experiments designed to gain further information on the existence and origin of the putative dehydroalanine residue in methylaspartase are underway.

Acknowledgements

We thank Roy Hartwell for N-terminal sequence determinations and the SERC for research grants GR/D-63400 and GR/E-44673 to D.G. Full details of the cloning and sequencing procedures will appear elsewhere.[23]

References

1. H. A. Barker, R. D. Smyth, R. M. Wilson and H. Weissbach, *J. Biol. Chem.,* 1959, **234**, 320.

2. V. R. Williams and J. G. Traynham, *Federation Proc.,* 1962, **21**, 247.

3. S. Ueda, K. Sato, and S. Shimizu, *J. Nutri. Sci. Vitaminol.,* 1982, **28**, 21.

4. M. F. Winkler, and V. R. Williams, *Biochim. Biophys. Acta.,* 1967, **146**, 287.

5. M. W. Hsiang and H. J. Bright, *J. Biol. Chem.,* 1967, **242**, 3079.

6. W. T. Wu and V. R. Williams, *J. Biol. Chem.,* 1968, **243**, 5644.

7. H. J. Bright, *J. Biol. Chem.,* 1964, **239**, 2307.

8. H. J. Bright, L. L. Ingraham and R. E. Lundin, *Biochim. Biophys. Acta,* 1964, **81**, 576.

9. K. R. Hanson and E. A. Havir, *Enzymes (3rd Ed.),* 1972, **7**, 75.

10. I. I. Nuiry, J. D. Hermes, P. M. Weiss, C. Chen and P. F. Cook, *Biochemistry,* 1984, **23**, 5168.

11. J. D. Hermes, P. M. Weiss and W. W. Cleland, *Biochemistry,* 1985, **24**, 2959.

12. S. C. Kim and F. M. Raushel, *Biochemistry,* 1986, **25**, 4744.

13. M. Akhtar, N. P. Botting, M. A. Cohen and D. Gani, *Tetrahedron,* 1987, **43**, 5899.

14. N. P. Botting, M. Akhtar, M. A. Cohen and D. Gani, *Biochemistry,* 1988, **27**, 2953.

15. N. P. Botting, M. A. Cohen, M. Akhtar and D. Gani, *Biochemistry,* 1988, **27**, 2956-2959.

16. N. P. Botting and D. Gani, *Biochemistry,* 1992, **31**, 1509.

17. N. P. Botting, A. Jackson and D. Gani, *J. Chem. Soc. Chem. Comm.,* 1989, 1583.

18. M. Young, N. P. Minton and W. L. Stadenbauer, *FEMS Microbiol. Rev.,* 1989, **63**, 301.

19. V. R. Williams and W. Y. Libano, *Biochim. Biophys. Acta.,* 1966, **118**, 144.

20. E. A. Havir and K. R. Hanson, *Biochemistry,* 1975, **14**, 1620.

21. I. L. Givot, T. A. Smith and R. H. Abeles, *J. Biol. Chem.,* 1969, **244**, 6341.

22. J. D. Hermes, C. A. Roeske, M. H. O'Leary and W. W. Cleland, *Biochemistry,* 1982, **21**, 5106.

23. S. Goda, N. P. Minton, N. P. Botting and D. Gani, *Biochemistry,* 1992, submitted.

Ligand Binding to Acetylcholine Receptors

V. B. Cockcroft[1], G. G. Lunt[2], and D. J. Osguthorpe[1]

[1] MOLECULAR GRAPHICS UNIT, UNIVERSITY OF BATH, CLAVERTON DOWN, BATH BA2 7AY, UK
[2] DEPARTMENT OF BIOCHEMISTRY, UNIVERSITY OF BATH, CLAVERTON DOWN, BATH BA2 7AY, UK

Introduction

Membrane associated receptors are extremely important to the functioning cells in many organisms. In multicellular organisms they allow for the control of a cells function by signals from other parts of the organism. One particularly important set of membrane bound receptors are those that are activated by neurotransmitters. There are 5 classes of transmitter, peptides, amino acids, monoamines, nucleotides and cholines. The receptors for these neurotransmitters have been extensively studied by molecular biology techniques and their amino acid sequences determined, both for many different receptors and for the same receptor in different species. Unfortunately, very little structural information for these receptors is available, primarily because of the current difficulty in determining the structure of membrane proteins. Understanding molecular recognition in these systems is therefore very difficult, and yet such an understanding would be extremely useful in the design of new ligands for modulating the activity of these systems.

The difficulty in determining the structure of membrane proteins is highlighted by the fact there is only one protein structure for a trans-membrane protein that has so far been determined to high resolution, the photosynthetic reaction centre.[1] Structures for bacteriorhodopsin,[2] a light-harvesting complex[3] and a porin[4] have been determined to reasonable resolution ($\approx$3-6Å) by electron microscopy. These are the only transmembrane structures known to-date. Only the reaction centre is a refined structure from X-ray crystallography in which the position of all atoms is known with reasonable certainty. Although the transmembrane helices are well-determined in the bacteriorhodopsin structure (from electron microscopy), the resolution is very uneven in different directions, varying from 3Å to 7Å.

Further, the positions of the loop residues connecting the helices is not known at all. The two structures also represent quite different classes of membrane proteins. In the reaction centre the membrane component is a small part of the overall reaction centre, most of which occurs outside the membrane. On the other hand, bacteriorhodopsin is almost entirely contained within the membrane, with the water exposed connecting loops being only a small part of the sequence. This makes it difficult to use these structures in looking for general principles of membrane protein structure.

One of the most highly studied neurotransmitters is the nicotinic acetylcholine receptor (nAChR). The nACh receptor allows cations to flow across the membrane into the cytoplasm of a neurone after binding the native ligand acetylcholine and other agonists. As such it is crucial to the function of the nervous system of many animals, including insects and thus provides a target for insecticidal compounds. Functionally, it is a multisubunit protein of five subunits, each of which has both an extracellular part, a transmembrane part and an intracellular part. The subunits form a complex which binds the receptor at some place on the extracellular part and this opens an ion-channel lined by a helix from each of the transmembrane parts of the five subunits. A number of other receptors have been identified, for example, the $GABA_A$ (γ-aminobutyric acid) receptor and the glycine receptor, which are similar functionally, i.e. they bind the agonist ligand which causes the opening of an integral ion-channel, although for the GABA and Gly receptors it is anions which are allowed to flow across the membrane at least in vertebrates. Extensive sequence data has been generated for the nAChR from different tissues and organisms,[5] and for the GABA[6] and Gly receptors.[7]

Analysis of the sequence data suggests the above receptors are members of a large, homologous set of proteins, the ligand-gated ion-channel (LGIC) superfamily.[8] The nACh receptor has the largest body of experimental data available for it, in particular for the receptor from *Torpedo* electric organ. This makes it the prototype for understanding the relation between structure and function within the LGIC class of receptors. Using the superfamily concept it should be possible to extrapolate ideas from studies on this receptor to all other LGIC receptors. Conversely, experimental information from other LGIC receptors can be used to confirm structure-relations in the nACh receptor.

However, even though the nAChR has been extensively studied and it is not difficult to obtain pure receptor from the electric organ of *Torpedo*, so far only very limited resolution structural information is available from electron microscsopy[9-11] of around 8 and 15 Å, which is not enough to determine the position of the backbone atoms of the protein. Thus, from this information alone it is impossible to rationalise in structural terms the activity of the various ligands or to rationally design ligands which are specific to a particular receptor.

Using the sequence data from molecular biology, a multiple amino acid sequence alignment of the LGIC family has been created from over seventy different subunits. These include subunits from receptors from different tissues, different species, different agonist ligands and receptors with different ion selectivity.[8] The sequence information shows that the subunits are homologous to each other, the lowest percentage identity being 19%. Furthermore, the extra-cellular region (the first ≈250 amino acids) and the trans-membrane region are significantly homologous, the percentage identity never less than 25%.

Two regions of the extracellular portion of the *Torpedo* nACh receptor have been suggested by biochemical studies to be involved in agonist binding. The first is the region around the paired cysteines 192-193 of the α-subunit which is labelled by various affinity ligands, including (4-N-maleimide)benzyltrimethylammonium (MBTA) and bromoacetylcholine,[12] after selective reduction of the disulphide bridge between them. In addition, a peptide extending from position 185 to 196 has been shown to bind α-bungarotoxin. [13] The second region is from position 125 to 147 of the α-subunit. This peptide contains a fifteen residue stretch of sequence termed the Cys-loop, so called because a disulphide bridge links Cysteine residues at positions 128 and 142,[12] (numbered from 1 to 15 in the rest of this article).

-Cys[128]-Ser-Ile-Asp-Val-Thr-Tyr-Phe-Pro-Phe-Asp-Gln-Gln-Asn-Cys[142]-

A synthetic peptide to this region was shown to interact with acetylcholine and α-bungarotoxin.[14] More recently, Madhok et al. [15] have reported that for the brain nACh receptor high-affinity binding of nicotine is specifically inhibited by antibodies raised to a peptide covering positions 130-139 of the Cys-loop of neuronal α-subunits.

Using the idea that all LGIC receptors have evolved from a single proto-receptor, we have concentrated on the Cys-loop region, which is the most conserved stretch of amino acid sequence in the diverse set of sequences of LGIC subunits known to date; 4 of its 15 residue positions are invariant. In contrast, only 14 residues are invariant in the entire N-terminal extracellular region of LGIC subunits, which is >200 residues long. Additionally, at position 11 of the Cys-loop an invariant aspartic acid residue occurs, which is one of only two invariant acidic residue positions present in the extracellular region of LGIC subunits. Therefore, this region must be important to the function of this receptor, either by being necessary for linking the binding of the agonist to the opening of the ion-channel or by being part of the binding site itself. The presence of the invariant aspartate suggests the latter as this would be a good candidate for the proposed anionic binding site, which would interact with the charged nitrogen that occurs in all LGIC agonists and most competitive antagonists. Some earlier proposals have been made[16, 17] about the structure of this region in the α subunit of the *Torpedo* nACh receptor but these did not include the mutual pairing of Cysteines 128-142. With the establishment of the concept of a LGIC superfamily we have been able to extend our analysis outwards from the nACh receptor and take into account the extensive biochemical, pharmacological and biophysical information for all known members of the receptor superfamily.[18]

The Cys-Loop Model

To create a specific model of the Cys-loop we used sequence conservation, spatial conservation and secondary structure prediction techniques to develop a model of its structure. From the sequence alignment a two residue periodicity in the conservation of hydrophobic positions suggested a β-strand structure, with an exposed hydrophilic face and buried hydrophobic face. The disulphide bridge between residues 1 and 15 of the loop thus requires a turn to occur around residues 8-9, where a conserved Pro residue occurs at position 9. By examining similar sequence turns occurring in known protein structures, in particular the requirement from the sequences for either phenylalanine or tyrosine at the preceding residue position, the starting conformation for this turn was chosen to be a type VIa β-turn.[19] This defined the conformation of the backbone of the Cys-loop, the side chain angles were taken from standard values for the β-sheet structure[20] Using energy minimisation and molecular dynamics techniques we refined the initial starting structure to create a low-energy conformation for the Cys-loop. Additionally, high temperature molecular dynamics was used to search for other nearby conformational energy minima but none were found.[18]

As a consistency check, the Cys-loop sequences for other members of the LGIC superfamily were put into the above conformation by mutation of the appropriate residues. All of the LGIC Cys-loop sequences had the same hydrophobic/hydrophilic separation, with the invariant Pro at position 9 preceded in all cases by an amino acid containing a benzene ring, i.e. Phe or Tyr (position 8) and the invariant Asp at position 11.[18] It was also possible to rationalise some aspects of the molecular recognition involved in ligand binding to these receptors to differences in sequence on the hydrophilic face. Thus, a hydrogen bonding residue (glutamine or histidine) occurs at position 13, whereas the residue at position 6 changes for each member of the LGIC superfamily, suggesting it is important in defining the selectivity of each LGIC member for its agonist. In the β-subunit of the GABA$_A$ receptor, it is an arginine residue, a lysine residue in the 48 kD subunit of the glycine receptor and a threonine residue in the α-subunit of the nACh receptor. The different size, charge and hydrogen bonding capability of these residues can be associated with a specific interaction with each of the different ligands that bind to that receptor, thus providing a method of discriminating between the various ligands. It is important to note that an interaction which provides great selectivity does not necessarily contribute greatly to the overall ligand binding energy. Thus, a solvated positive and negative charge when brought close together and desolvated have a near zero overall energy, but if either of the charges are removed or changed in sign the energy of the desolvated, binding complex in principle becomes very large, i.e. the two entities now have a large, unfavourable energy for complex creation.

Extended Binding Site Model

So far, the modelling has created an explicit, atomic-level detail model of this small section of the protein that we propose forms part of the binding site for agonists. However, there is other data available, mainly from biochemical studies, which can be used to extend this model. In particular, the region around Cys 192-193 has long been identified as part of the binding site, from studies using reagents specific for the binding site that alkylate these cysteines in the *Torpedo* nACh receptor. For these reagents the active part of the molecule is in a position equivalent to the bromomethyl group of bromoacetylcholine and is, therefore, probing for residues in a limited region of space around the binding site. We have modelled this region as being close to the agonist binding site and part of a binding cleft. This region is important for toxin binding, such as bungarotoxin, but the large size of these toxins means they only need to bind close to the active site to occlude it.

Additional information on the binding site cleft can be provided by pharmacophore mapping using the extensive range of agonists and antagonists known for the LGIC family. By superimposing the differing agonists and antagonists for all LGIC members that are known to act at the native agonist binding site, a spatial map of the binding site cleft can be constructed. This was used in the positioning of other components of the agonist binding site around the Cys-loop hydrophilic surface. The superimposition was accomplished using a 3 point pharmacophore model for agonist and competitive antagonist binding, which includes the charged nitrogen of all agonists, and the electronegative and electropositive atoms of the polarised π-electron system found close to the charged nitrogen, which is also found in all agonists and some antagonists.[18]

Trans-membrane Helices Modelling

From hydrophobicity profiles calculated from the sequence data, 4 regions of the nAChR (M1-M4) have been identified as possible transmembrane helices,[21] and 1 region (MA) that has been identified as a highly charged amphiphilic helix.[22] Helices are the most likely secondary structure to be found in a low-dielectric medium as the helical structure allows the backbone hydrogen bonding groups to interact favourably and yet it only requires short range contacts. This has been demonstrated experimentally by helix-coil transition studies on various amino acid polymers in differing solvents, which show that in low-dielectric solvents the helical conformation dominates. Hence the general assumption that the most likely secondary structure in a membrane will be helical.

Previously, models of the amphiphilic helix showed that all the charged residue side-chains could be aligned down one side of the helix surface and because of this it has been suggested that it is the helix which lines the ion-channel.[22, 23] Recent work by Numa[24, 25] demonstrated that partial ion-channel activity still existed even after the deletion (by molecular biological techniques) of the MA helix region. Having removed MA as a contender for the ion-channel lining helix, attention now has focussed on the M2 helix. This contains a number of Ser and Thr residues which can be lined up along one side of the helix surface. Thus, the hydroxyl side chains of these residues could replace, or interact with, the solvent shell surrounding the ion as it is passing through the channel. Also, novel peptides have been synthesised based on the M2 sequence which have been shown to act as ion-channels.[26]

In the creation of a model for the transmembrane region, the major question is how do the other helices pack around M2 so that when the subunits are put together each M2 helix is lining the ion-channel. A suitable position for the M4 helix is lying on the outside of the helix bundles and making no contacts with the helices of adjacent subunits. This puts the least structural constraints on this helix and the sequence alignments show that this is the least conserved helix. Further, Numa has replaced this helix with a hydrophobic helix from the insulin receptor and retained activity,[25] which again indicates there are few structural constraints on this helix.

Extra-cellular domain Modelling

We have created some models of a small portion of the extra-cellular domain, the agonist binding site, and for the trans-membrane domain. We would like to construct a model of the whole extra-cellular domain which could be incorporated with the trans-membrane domain to produce a model which could be used to understand the overall function of the receptor, i.e. how binding the agonist leads to the opening of the ion-channel. However, using standard techniques no homology has been found between the extra-cellular domain and any protein of known structure. It does appear that the extra-cellular domain is entirely external to the membrane, as shown by electron micrographs, it is highly likely that its structure is determined by the same rules as for normal globular proteins, for which we have a large body of structural data.

Using a method we have recently developed for scanning the protein structure database with sequences of unknown structure and no apparent homology,[27] we have identified the protein, of known structure, pyrophosphatase (PPase) as having a high degree of similarity to the extra-cellular domain of the nACh receptor. A pair-wise alignment of these two sequences was prompted by this result and this gave an 18% identity conservation between them. This low level of sequence homology has in other modelling work[28, 29] been used to generate structures by homology modelling which have been shown to be overall correct by subsequent X-ray studies.

Conclusions

Understanding molecular recognition in systems for which little structural data is available, and little hope of structural information being available in the near future, is fraught with difficulty. Rather than doing nothing, we can make suggestions about aspects of recognition from the information that is available from molecular biology. The main benefit of the construction of explicit atomic models, besides giving insights into receptor structure and function, is that it allows specific experiments to be designed and prioritised to test the model and to comprehend and verify current experimental data. This modelling is a first step in the understanding of the structure-activity relationships of receptor ligands, whilst taking into consideration the receptor sites themselves.

Acknowledgements

We thank the SERC and Shell Research (UK) Ltd. for financial support.

References

1. J. Deisenhofer, O. Epp, K. Mikki, R. Huber, and H. Michel, *Nature*, **318**, 618 624 (1985).

2. R. Henderson, J.M. Baldwin, T.A. Ceska, F. Zemlin, E. Beckmann, and K.H. Downing, *J. Mol. Biol.*, **213** (1990). in press.

3. W. Kuhlbrandt and Da Neng Wang, *Nature*, **350**, 130 134 (1991).

4. B.K. Jap, P.J. Walian, and K. Gehring, *Nature*, **350**, 167 170 (1991).

5. M. Noda, H. Takahashi, T. Tanabe, M. Toyosato, Y. Furutani, T. Hirose, M. Asai, S. Inayama, T. Miyata, and S. Numa, *Nature*, **299**, 793 797 (1982).

6. P. R. Schofield, M. G. Darlison, N. Fujita, D. Burt, F. A. Stephenson, H. Rodriguez, L. M. Rhee, J. Ramachandran, V. Reale, T. A. Glencorse, P. H. Seeburg, and E. A. Barnard, *Nature*, **328**, 221 227 (1987).

7. G. Grenningloh, A. Rienitz, B. Schmitt, C. Methfessel, K. Beyrether, E. D. Gundelfinger, and H. Betz, *Nature*, **328**, 215 220 (1987).

8. V. B. Cockcroft, D. J. Osguthorpe, A. F. Friday, E. A. Barnard, and G. G. Lunt, *Molecular Neurobiology*, **4**, 129 169 (1990).

9. C. Toyoshima and N. Unwin, *Nature*, **336**, 247 250 (1988).

10. N. Unwin, *Neuron*, **3**, 655 676 (1989).

11. M. Mitra, M.P. McCarthy, and R.M. Stroud, *J. Cell. Biol.*, **109**, 755 774 (1989).

12. P. N. Kao and A. Karlin, *J. Biol. Chem.*, **261**, 8085 8088 (1986).

13. D. Neumann, D. Barchan, M. Fridkin, and S. Fuchs, *Proc. Natl. Acad. Sci. USA*, **83**, 9250 9253 (1986).

14. D. J. McCormick and M. Z. Atassi, *Biochem. J.*, **224**, 995 1000 (1984).

15. T. C. Madhok, C. C. Chao, S. Matta, A. Hong, and B. M. Sharp, *Biochem. Biophys. Res. Com.*, **165**, 151 157 (1989).

16. L. Smart, H. Meyers, R. Hilgenfeld, W. Saenger, and A. Maelicke, *FEBS Lett.*, **178**, 64 68 (1984).

17. W.H.M.L. Luyten, *J. Neuroscience Res.*, **16**, 51 73 (1986).

18. V.B. Cockcroft, D.J. Osguthorpe, E. Barnard, and G.G. Lunt, *Proteins*, **8**, 386 397 (1990).

19. P. Y. Chou and G. D. Fasman, *J. Mol. Biol.*, **115**, 135 175 (1977).

20. M.J. Sutcliffe, F.R.F. Hayes, and T.L. Blundell, *Protein Eng.*, **1**, 385 392 (1987).

21. T. Claudio, M. Ballivet, J. Patrick, and S. Heinemann, *Proc. Natl. Acad. Sci. USA*, **80**, 1111 1115 (1983).

22. J. Finer-Moore and R. M. Stroud, *Proc. Natl. Acad. Sci.*, **81**, 155 159 (1984).

23. R. M. Stroud and J. Finer-Moore, *Ann. Rev. Cell Biol.*, **1**, 317 351 (1985).

24. M. Mishina, T. Tobimatsu, K. Imoto, K. Tanaka, Y. Fulita, Y. Fukuda, T. Hirose, S. Inayama, T. Takahashi, M. Kuno, and S. Numa, *Nature*, **313**, 364 369 (1985).

25. T. Tobimatsu, Y. Fujita, K. Fukuda, K. Tanaka, Y. Mori, T. Konno, M. Mishina, and S. Numa, *FEBS. Lett.*, **222**, 56 62 (1987).

26. J.D. Lear, Z.R. Wasserman, and W.F. DeGrado, *Science*, **240**, 1177 1181 (1988).

27. V.B. Cockcroft and D.J. Osguthorpe, *FEBS Lett.*, **293**, 149 152 (1991).

28. L.H. Pearl and W.R. Taylor, *Nature*, **329**, 351 354 (1987).

29. T. Blundell and L. Pearl, *Nature*, **337**, 596 597 (1989).

Peptides and Amino Acids of Medicinal Importance

R. A. August, A. N. Bowler, P. M. Doyle, X. J. Durand,
P. Hudhomme, J. A. Khan, C. M. Moody and D. W. Young

SCHOOL OF CHEMISTRY AND MOLECULAR SCIENCES, UNIVERSITY OF SUSSEX, FALMER,
BRIGHTON, BNI 9QJ, UK

1   INTRODUCTION

This lecture reports studies on peptides and amino acid
derivatives which are important in four distinct
therapeutic areas. The link between the four projects is
in the synthetic methods which have been developed to
effect their solution.

2   PROTEINS INVOLVED IN THE BLOOD CLOTTING CASCADE

The first project concerns peptides in the so-called
coagulation cascade. This cascade involves activation of
a series of inactive plasma proteins by minor
proteolysis.[1] Activation converts a precursor protein to
a serine protease which is, in turn, responsible for
activation of the next protein in the cascade. The
cascade eventually leads to fibrin formation and
generation of an insoluble clot.

The various protein factors in the coagulation
cascade all have an amino terminal domain which contains
nine to twelve γ-carboxyglutamic acid (Gla) residues (2)
and their formation is dependent on the presence of
vitamin K (5). This domain appears to bind platelet
phospholipid in the presence of calcium ions, the
malonate units in the Gla residues being important in
chelating these ions. The Gla residues are derived by
post-translational γ-carboxylation of glutamic acid
residues and inhibition of this carboxylation is of
relevance for the development of anti-thrombotic drugs.

The carboxylation has been shown to occur as outlined
in Scheme 1 below and involves stoichiometric quantities
of the cofactor, vitamin K-hydroquinone (3), CO_2 and O_2.
The products of the reaction are γ-carboxyglutamic acid
(2), vitamin K epoxide (4) and H_2O. The hydroquinone
cofactor (3) is regenerated from the epoxide (4) via the
vitamin (5). The anti-coagulant warfarin (6), which has
some structural resemblance to the vitamin, acts by

inhibiting the enzyme(s) responsible for the reductive
regeneration of the cofactor.

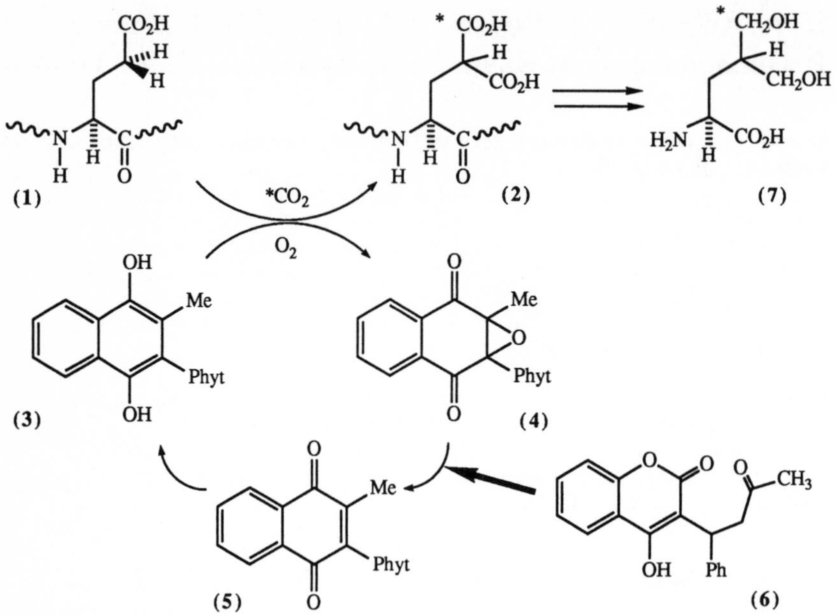

Scheme 1

Our first interest was to determine the overall
stereochemistry of the carboxylation process. Marquet *et
al.*[2,3] had shown that the *4-pro-S* hydrogen of glutamic
acid is lost in the process, and so we opted to
investigate the stereochemistry of the addition of the γ-
carboxyl group. The diastereotopic carboxyl groups were
known to absorb at different chemical shift values in the
^{13}C-nmr spectrum[4] and so, if absolute stereochemistry
could be assigned to these absorptions, the ^{13}C-nmr
spectrum of the product derived from incubating a small
peptide with the enzyme and [^{13}C]-CO_2 would allow the
overall stereochemistry of the carboxylation to be
determined.

The most direct way to assign stereochemistry to the
absorptions in the ^{13}C-nmr spectrum would be to effect a
synthesis of stereospecifically [^{13}C]-labelled γ-
carboxyglutamic acid. The method of synthesis would then
define the stereochemistry of labelling. The fact that
this synthesis would involve transformations on
enolisable malonate esters made stereospecifically
labelled γ-carboxyglutamic acid an impracticable goal and
we therefore chose labelled dihydroxyleucine (7) as our
revised target. This compound had been obtained[5,6] from
Gla-containing peptide by reduction with diborane and
proteolysis. Dihydroxyleucine has a stable stereocentre

at C-4 and, since labelling with deuterium is as effective as labelling with ^{13}C in allowing the ^{13}C-nmr spectrum to be assigned, synthesis of a sample of dihydroxyleucine labelled with deuterium in only one of the two diastereotopic CH_2OH groups would allow the stereospecificity of the enzymic carboxylation to be assessed.

We chose (2S)-pyroglutamic acid **(8)** as the starting point for our synthesis, since the ring might allow the centre at C-2 to induce stereochemistry at C-4   Further, the ring structure will allow stereochemistry to be defined using n.O.e. experiments, before conversion to acyclic precursors of labelled dihydroxyleucine.   We therefore converted (2S)-pyroglutamic acid **(8)** to the N-*tert*-butoxycarbonyl   *tert*-butyl   ester   **(10)**   by esterification followed by reaction with $(Boc)_2O$ and DMAP.   Reaction of this compound with Bredereck's reagent[7] then gave an excellent yield of the enaminone **(11)**.   Acid hydrolysis to the aldehyde and reduction *in situ* with $NaB(CN)H_3$ gave a mixture of the *trans-* and *cis-* alcohols **(12)** and **(13)** in a ratio of 70% to 30%.   Since we expected the intermediate *trans*-aldehyde to be thermodynamically more stable than the *cis*-aldehyde, we assumed that the *trans*-alcohol would be the predominant isomer.   This was confirmed by the presence of a n.O.e. between H-2 and H-*3-pro-S* and H-*3-pro-R* and H-4 in the *trans*-alcohol **(12)** and a n.O.e. between H-2 and H-*3-pro-S* and H-*3-pro-S* and H-4 in the *cis*-alcohol **(13)**.

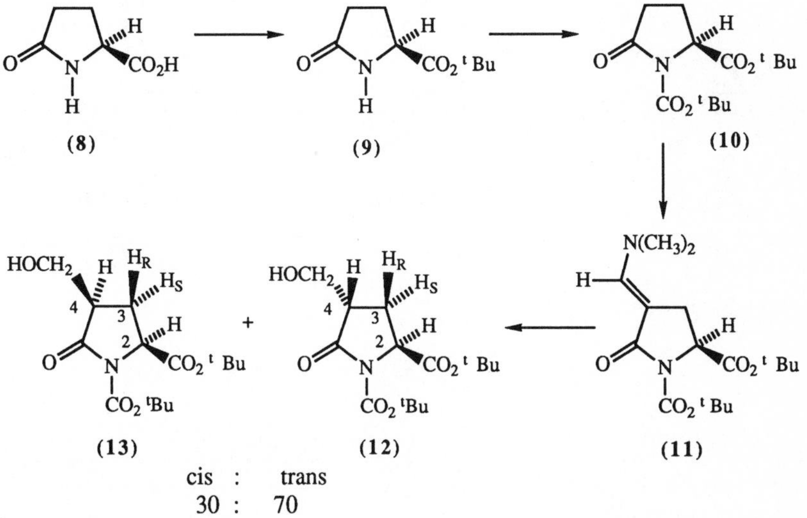

Scheme 2

The alcohols could be separated by column chromatography but, since they eluted very close together, separation was tedious.   We had evidently succeeded in achieving stereoselective synthesis by relying on thermodynamic control, and so we decided to investigate kinetic control as a means of obtaining a truly stereospecific synthesis of the *cis*-isomer. We chose *trans*-4-hydroxy-(2S)-proline (**14**) as our starting material, and converted this, *via* the N-*tert*-butoxycarbonyl-ketone (**16**), esterification and Wittig reaction to the exomethylene compound (**17**) as shown in Scheme 3. This was now treated with disiamylborane followed by H_2O_2 to give the alcohol (**18**). The well-known problem of conformational isomerism of N-acylproline derivatives[8] complicated assessment of the stereochemistry using nmr spectroscopic methods but, when the alcohol was converted to the TBDPS ether and then oxidised to the (2S)-pyroglutamic acid derivative (**19**), it was evident that the product was a mixture in which the *cis*-isomer (**19**) predominated to the extent of 80%. When $BH_3.Me_2S$ was used in the hydroboration step, a 50:50 mixture of the *cis*- and *trans*-isomers (**19**) and (**20**) was obtained.

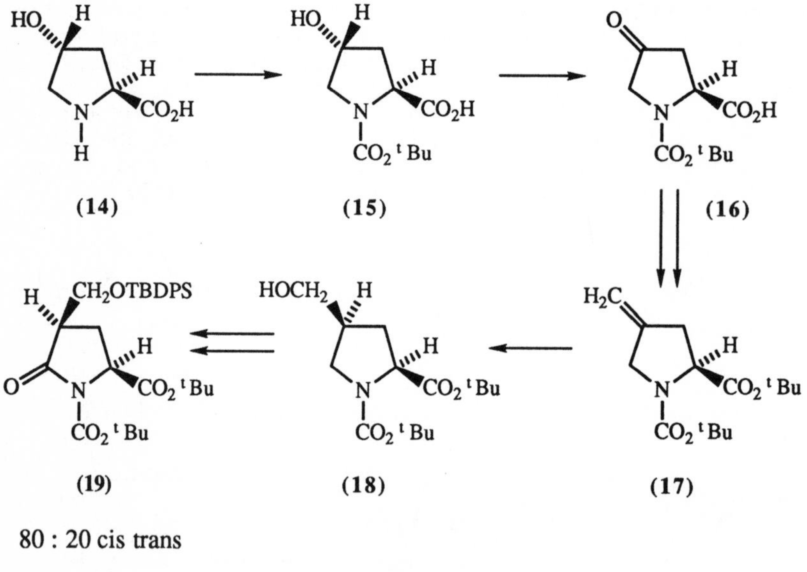

(**14**)   (**15**)   (**16**)

(**19**)   (**18**)   (**17**)

80 : 20 cis trans

Scheme 3

Although we had still achieved only stereoselective synthesis, the *cis*- and *trans*- TBDPS ethers (**19**) and (**20**) were readily separated in good yield by column chromatography and the stereochemical assignments were confirmed by n.o.e. experiments. The mixed alcohols from the synthesis outlined in Scheme 2 were also converted to the TBDPS ethers and separated and so we could prepare large amounts of either the *trans*-isomer (**20**) or the *cis*-isomer (**19**) by use of the two syntheses.

Having achieved synthesis of two protected alcohols of well-defined stereochemistry, the next step was to open the pyroglutamate ring to obtain a substituted glutamic acid derivative. The γ-carboxyl group would then be converted to the required C^2H_2OH group without the α-ester being affected. Attempts at ring-opening using esters other than *tert*-butyl had resulted in transesterification or hydrolysis of the α-ester so that the desired regioselectivity was not possible. The α-*tert*-butyl ester, however, proved ideal for our purposes. Initial attempts at ring-opening the α-tert-butyl ester **(19)** using Et$_3$N/ MeOH resulted in the desired α-*tert*-butyl γ-methyl diester but this was accompanied by some epimerisation at the γ-centre. Aqueous LiOH in THF did result in ring-opening but the basic conditions also resulted in elimination of the TBDPSO group to yield the exomethylene derivative **(21)**. It was evident that we required a weaker base and stronger nucleophile to effect ring-opening without elimination and so aqueous LiOOH in THF was employed. This reagent gave a reasonable yield of the desired acid **(22)**. Conversion to the mixed anhydride and reduction with NaB^{2}H$_4$ currently shows promise and hydrolysis will then yield (2S,4S)-[5,5-^{2}H$_2$]-5,5-dihydroxyleucine, our target compound for study of the γ-carboxylation of glutamic acid.

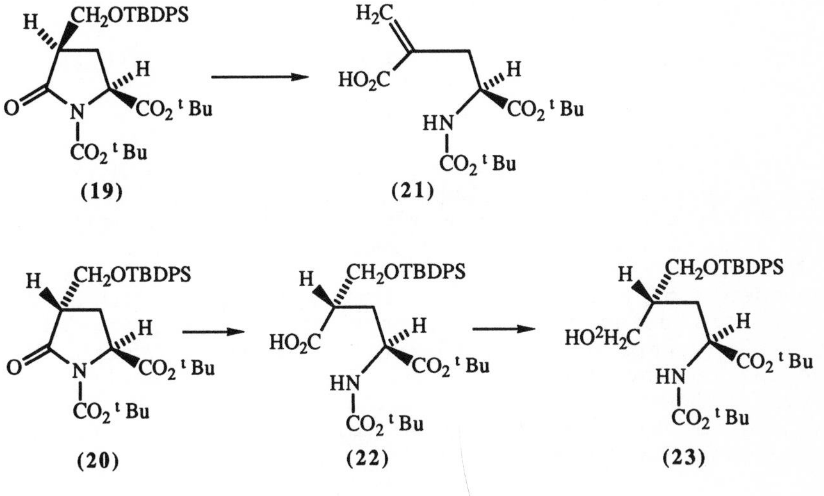

Scheme 4

One of the interests in glutamate γ-carboxylase is the possibility that inhibitors of the enzyme might be useful anti-thrombotic drugs. Such inhibitors may be modelled either on the cofactor **(3)** or the substrate **(1)** of the enzyme. Since smaller peptides than those in the blood clotting cascade have been shown to be substrates for the biological carboxylation reaction,[9] and indeed, even protected glutamic acids have been shown to be substrates, useful inhibitors need not be peptides. Although the mechanism of carboxylation is far from clear, it occurred to us that, if the γ-enolate of a glutamic acid residue were involved, then the transition state might resemble **(24)** below. Thus compounds of general structure **(25)** might be useful transition state inhibitors.

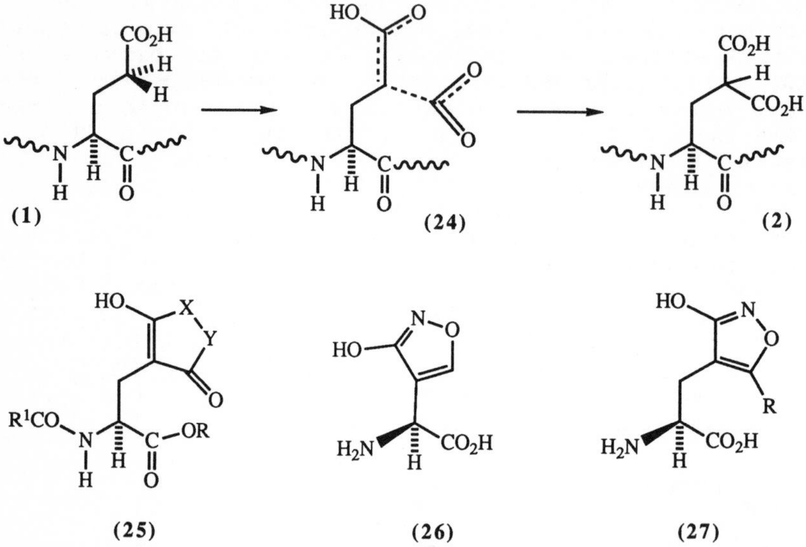

(1)   **(24)**   **(2)**

(25)   **(26)**   **(27)**

Scheme 4

On searching for such compounds in the literature, we noted that among the compounds prepared following the discovery of ibotenic acid **(26)**, the active principle of the mushroom *Amanita muscaria*, were compounds such as **(27)** which acted upon the AMPA receptors. Since the syntheses of these compounds involved routine amino acid synthesis using a separate heterocyclic building block for each compound and resulted in racemic products,[10-17] we decided to develop a more flexible synthesis which would allow access to a series of enantiomerically pure analogues of the AMPA agonist **(27)**.[18] This took us into our second therapeutic area.

3. EXCITATORY AMINO ACIDS

Having prepared an aldehyde intermediate on hydrolysis of the enaminone **(11)**, it seemed to us that reaction of such a compound with α-nucleophiles such as substituted hydrazines or hydroxylamine might yield an intermediate which would undergo further reaction *in situ*. The product would be a protected analogue of the agonist **(27)**. We therefore prepared the benzyl N-carbobenzyloxy enaminone **(28)**,[19,20] hydrolysed it until the uv spectrum indicated complete conversion to the aldehyde and treated this *in situ* either at pH 1 or buffered at pH5 with a variety of α-nucleophiles.

Reaction with methylhydrazine at pH 5 initially showed the presence of two spots on TLC but one was converted to the other within 15 hours. The final product had lost the absorption at 1790cm^{-1} in the ir spectrum due to the protected pyroglutamate and had nmr spectra compatible with the structure **(31, X=NCH₃)**. This appeared to exist as the enol tautomer but acetylation gave the acetamide **(32, X=NCH₃)** rather than the enolether.

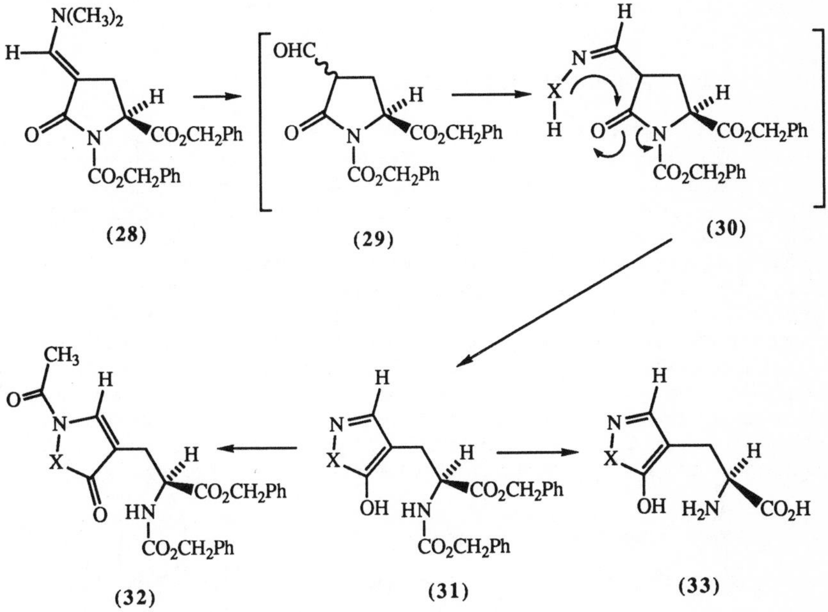

Scheme 6

Reaction with hydrazine, phenylhydrazine, and *para*-nitrophenylhydrazine yielded the "ring switched" products **(31, X=NH)**, **(31, X=NPh)** and **(31, X=N-Ph-pNO₂)** respectively. The compounds were all optically active and no deuterium was incorporated when the reaction was conducted in CH₃O²H so that the process had not affected the chirality of the α-centre. When the reaction with *para*-nitrophenylhydrazine was conducted at pH 1, the *para*-nitrophenylhydrazone was the only isolable product, and use of 2,4-dinitrophenylhydrazine at either pH 1 or pH 5 gave only the 2,4-dinitrophenylhydrazone.

Reaction of the intermediate aldehyde **(29)** with hydroxylamine gave the isoxazole **(31, X=O)** which was present as the enol tautomer, and which, unlike the corresponding pyrazoles **(31, X=NR)**, gave the enol acetate on acetylation. The isoxazole **(31, X=O)** appeared to be unstable on standing, and it is interesting that, subsequent to our preliminary communication of these results,[18] a novel amino acid antibiotic TAN-950 A was shown to have the structure **(33, X=O)**.[21] This equilibrated to a mixture of stereoisomeric pyroglutamic acid oximes by a mechanism which is the reverse of our "ring switching" mechanism.

So far, we have deprotected two of our protected heterocyclic amino acids by hydrogenation in glacial acetic acid using Adams catalyst. The products **(33, X=NH)** and **(33, X=NCH3)** demonstrated an inhibitory effect on ibotenate stimulated phosphoinositol response.

4. DIHYDROFOLATE REDUCTASE

Our involvement in the final two therapeutic areas which we shall discuss started with a long-standing interest in the enzyme dihydrofolate reductase which is a target enzyme for cancer drugs such as methotrexate **(34)**. We had earlier shown that the vitamin folic acid **(35)** was reduced to the coenzyme tetrahydrofolic acid **(36)** by this enzyme with addition of hydrogen from the *4-pro-R* hydrogen of NADPH to C-6 and C-7 of the pteridine ring from the opposite face to the face of methotrexate which had been shown to come into contact with NADPH in a ternary complex.[22,23] Subsequently, Feeney *et al*.[24] showed that the methyl groups in two leucine residues (Leu-19 and Leu-27) in the enzyme came within n.O.e. distance at H-7 of methotrexate bound at the active site.

(34)   (35)   (36)

It seemed to us that if samples of leucine labelled in just one of the two diastereotopic methyl groups could be made in sufficient quantities to be incorporated into the enzyme by biological means, then the stereochemistry of these interactions would be defined. We therefore, devised a synthesis of leucine labelled in one of the two diastereotopic methyl groups starting from the enaminone **(11)**. This will allow nmr spectroscopic techniques to be used to investigate protein-protein and protein-small molecule interactions in general. Previous syntheses of labelled leucine[25-30] were not totally stereospecific, and it was essential that our synthesis be stereospecific and efficient, if sufficient quantities of the amino acid were to be obtained to be incorporated into protein.

The enaminone **(11)** was reduced to the exomethylene derivative **(37)** using DIBAL, and this was found to be catalytically reduced with total asymmetric induction to yield the *cis*-4-methyl derivative **(39)**. In early work, the DIBAL reduction yielded a mixture of diastereoisomeric amines **(38)** as byproduct and yet we found that this mixture of isomers could be catalytically reduced to the pure *cis*-4-methyl compound **(39)**. We therefore investigated direct catalytic reduction of the enaminone **(11)** and obtained a 78% yield of the pure *cis* product **(39)**.

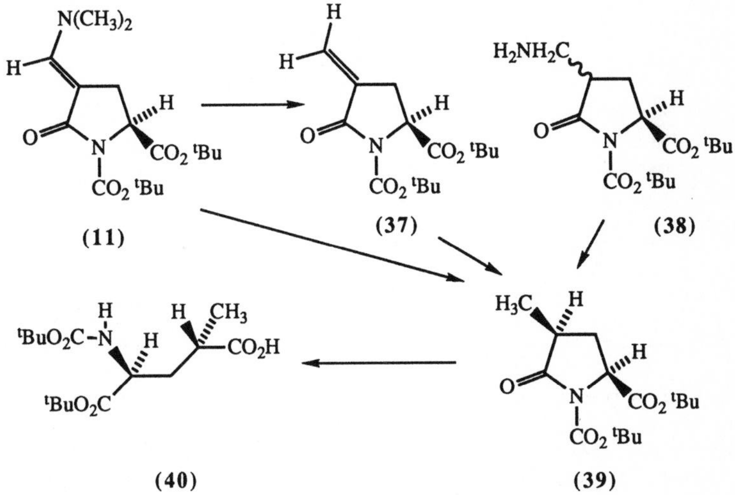

Scheme 7

The pyroglutamate template had, therefore, given us complete asymmetric induction, and we now effected ring-opening using aq. LiOH in THF to obtain the acid **(40)** in 94% yield. The acid was now converted to the labelled alcohol **(41)** in an overall yield of 75% by conversion to the mixed anhydride and reduction with NaB^2H_4. The synthesis was completed by conversion of the alcohol **(41)** to the iodide using methyltriphenoxyphosphonium iodide in HMPA, reduction with $NaB(CN)^2H_3$ and hydrolysis in 6M aqeous HCl.

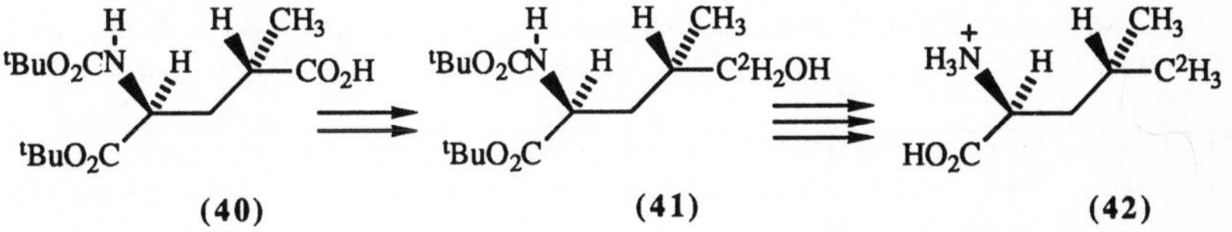

Scheme 8

Part of the ^{13}C-nmr spectrum of the resultant (2S,4R)-[5,5,5-^{2}H$_3$]-leucine is shown in Figure 1 and a two-dimensional ^{13}C-^{1}H shift correlation (shown in Figure 2) allowed the shifts of the diastereotopic methyl groups in the ^{1}H-nmr spectrum to be assigned . This synthesis has now been completed on a large scale and the product has been used as the only source of leucine for an auxotrophic strain of *Lactobacillus casei* by Dr. J. Feeney and his colleagues at The National Institute for Medical Research, Mill Hill. The enzyme is currently in the latter stages of purification.

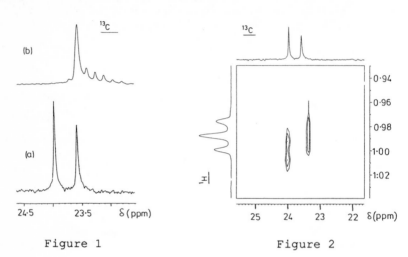

Figure 1   Figure 2

^{13}C-nmr spectra in ^{2}HCl of : ^{13}C-^{1}H correlation of
   (a) (2S)-leucine   (2S)-leucine in ^{2}HCl
   (b) (2S,4R)-[5,5,5,^{2}H$_3$]-
      leucine

Since the alcohol **(41)** might serve as a synthon for samples of leucine with fluorine in the *4-pro-R* methyl group and enzymes containing this label will allow biological interactions to be probed using ^{19}F-nmr spectroscopy, we decided to synthesise (2S,4S)-5-fluoroleucine **(47)**. This has now been achieved on a small scale. Initially, reaction of the alcohol **(41)** with DAST led to cyclisation to the proline derivative **(43)** and so double protection on nitrogen was required to prevent the intramolecular substitution reaction. This was achieved by first protecting the alcohol **(41)** using TBDMS-chloride. The ether **(44)** was then protected using (Boc)$_2$O, DMAP and Et$_3$N. The resultant product **(45)** was deprotected with (Bu)$_4$N$^+$ F$^-$ and treatment of the alcohol **(46)** with DAST followed by deprotection gave the required product **(47)**.

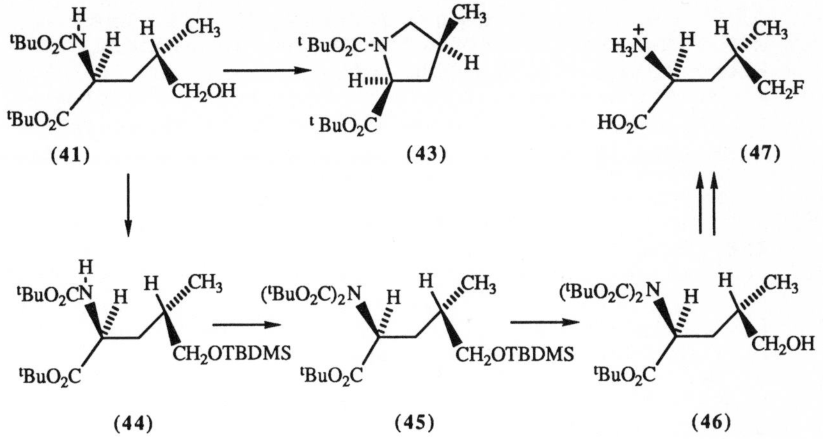

Scheme 9

5. LEUCINE ZIPPER PROTEINS

There is an important class of DNA-modifying proteins which are dimeric and contain a helical region with a leucine residue at every seventh amino acid in the helix.[31] This results in a series of four hydrophobic leucine residues occurring at every second turn on the same side of the helix in each monomer. This "leucine zipper" region holds the monomeric units together by hydrophobic interactions and incorporation of our stereospecifically labelled samples of leucine (42) into these proteins will allow the dimerisation to be studied in detail using nmr spectroscopy. We are collaborating with Dr Julian Burke in our own University on such a protein from yeast and have achieved overexpression in *Escherichia coli*. The scale-up of this construct is currently being investigated. Further collaboration with Dr. Andrew Lane at Mill Hill on two mammalian zipper proteins is also in hand. Having proteins from three sources will allow us to study both homo- and hetero-dimerisation processes.

REFERENCES

1. E.W. Davie, K. Fujikawa and W. Kisiel, *Biochemistry*, 1991, **30**, 10363.
2. P. Decottignies-Le Marechal, C. Ducrocq, A. Marquet and R. Azerad, *J. Biol. Chem.*, 1984, **259**, 15010.
3. P. Decottignies-Le Marechal, C. Ducrocq, A. Marquet and R. Azerad, *C. R. Acad. Sc. Paris*, 1984, **298**, (Ser. II), 343.
4. R. Sperling, B.C. Furie, M. Blumenstein, B. Keyt and B. Furie, *J. Biol. Chem.*, 1978, **253**, 3898.
5. T.H. Zytkovicz and G.L. Nelsestuen, *J. Biol. Chem.*, 1975, **250**, 2968.

6. C. Vermeer, J.W.P. Govers-Riemslag, B.A.M. Soute,
    M.J. Lindhout, J. Kop and H.C. Hemker, *Biochim.
    Biophys. Acta*, 1978, **538**, 521.
7. H. Bredereck, G. Simchen, S. Rebsdat, W. Kantlehner,
    P. Horn, R. Wahl, H. Hoffmann and P. Grieshaber,
    Chem. Ber., 1968, **101**, 41.
8. J. Hondrelis, G. Lonergan, S. Voliotis and
    J. Matsoukas, *Tetrahedron*, 1990, **46**, 565.
9. J.J. McTigue and J.W. Suttie, *J. Biol. Chem.*, 1983,
    258, 12129.
10. J.J. Hansen and P. Krogsgaard-Larsen, *J. Chem. Soc.*,
    Perkin Trans. 1, 1980, 1826.
11. P. Krogsgaard-Larsen, L.Brehm, J.S. Johansen,
    P. Vinzents, J. Lauridsen and D.R. Curtis,
    J. Med. Chem., 1985, **28**, 673.
12. K. Sirakawa, O. Aki, S. Tsushima and K. Konishi,
    Chem. Pharm. Bull., 1966, **14**, 89.
13. Y. Kishida, T. Hiraoka, J. Ide and A. Terada,
    Chem. Pharm. Bull., 1967, **15**, 1025.
14. N. Nakamura, *Chem. Pharm. Bull.*, 1971, **19**, 46.
15. A.R. Gagneux, F. Hafliger, R. Meier and C.H. Eugster,
    Tetrahedron Letters, 1965, 2081.
16. T. Honore and J. Lauridsen, *Acta Chem. Scand.*,
    Sect. B, 1980, **34**, 235.
17. J. Lauridsen, T. Honore and P. Krogsgaard-Larsen,
    J. Med. Chem., 1985, **28**, 668.
18. A.N. Bowler, P.M. Doyle and D.W. Young,
    J. Chem. Soc., Chem. Commun., 1991, 314.
19. H. Gibian and E. Klieger, *Annalen*, 1961, **640**, 145.
20. S. Danishefsky, E. Berman, L.A. Clizbe and M. Hirama,
    J. Am. Chem. Soc., 1979, **101**, 4385.
21. S. Tsubotani, Y. Funabashi, M. Takamoto, S. Hakoda
    and S. Harada, *Tetrahedron*, 1991, **47**, 8079.
22. P.A. Charlton, D.W. Young, B. Birdsall, J. Feeney and
    G.C.K.Roberts, *J. Chem. Soc., Chem. Commun.*,
    1979, 922.
23. P.A. Charlton, D.W. Young, B. Birdsall, J. Feeney and
    G.C.K.Roberts, *J. Chem. Soc., Perkin Trans. 1*,
    1985, 1349.
24 B. Birdsall, J. Feeney, S.J.B. Tendler, S.J. Hammond
    and G.C.K.Roberts, *Biochemistry*, 1989, **28**, 2297.
25. R. Cardillo, C. Fuganti, D. Ghiringhelli,
    P. Grasselli and G. Gatti, *J. Chem. Soc.*,
    Chem. Commun., 1977, 474.
26. C. Fuganti, P. Grasselli and G. Pedrocchi-Fantoni,
    Tetrahedron Letters, 1979, 2453.
27. P. Anastasis, I. Freer, K.H. Overton, D. Picken,
    D.S. Rycroft and S.B. Singh, *J. Chem. Soc.*,
    Perkin Trans. 1, 1987, 2427.
28. D.J. Aberhart and B.H. Weiller, *J. Labelled Compounds
    and Radiopharm.*, 1983, **20**, 663.
29. S.R. Sylvester and C.M. Stevens, *Biochemistry*,
    1979, **18**, 4529.
30. S.R. Sylvester, S.Y. Lan and C.M. Stevens,
    Biochemistry, 1981, **20**, 5609.
31. S.L. McKnight, *Scientific American*, 1991 (April), 32.

Molecular Recognition and Drug Design: The Structural Basis of Specificity of Human and Mouse Renins Defined by *X*-Ray Analyses of Peptide Inhibitor Complexes

V. Dhanaraj[1], C. Dealwis[1], C. Frazao[1], M. Badasso[1],
I. J. Tickle[1], J. B. Cooper[1], M. Newman[1], C. Aguilar[1],
S. P. Wood[1], T. L. Blundell[1], P. M. Hobart[2], K. F.
Geoghegan[2], M. J. Ammirati[2], D. E. Danley[2], B. A.
O'Connor[2] and D. J. Hoover[3]

[1] ICRF UNIT OF STRUCTURAL MOLECULAR BIOLOGY, DEPARTMENT OF
CRYSTALLOGRAPHY, BIRKBECK COLLEGE, LONDON WC1E 7HX, UK
[2] DEPARTMENT OF MOLECULAR GENETICS AND PROTEIN CHEMISTRY, CENTRAL
RESEARCH, PFIZER INC, GROTON CT 06340, USA
[3] DEPARTMENT OF MEDICINAL CHEMISTRY, CENTRAL RESEARCH, PFIZER INC,
GROTON, CT 06340, USA

1 INTRODUCTION

Molecular recognition plays an important role in enzyme catalysis. Enzymes have evolved to recognise the transition states of substrates both at the reaction centre and at the specificity subsites. Clues as to the factors involved in such recognition processes can be obtained from the study of the three-dimensional structures of enzyme complexes with transition isosteres. This is very well illustrated by high resolution X-ray structures of peptide complexes of the important drug target, renin.

Renin[1] is the key enzyme of the renin-angiotensin cascade which regulates the formation of angiotensin II, a potent pressor and aldosterogenic substance controlling vascular tone, fluid volume and sodium excretion. Because of the unique specificity of renin, its inhibition is widely expected to provide selective therapy for hypertension, congestive heart failure and associated degenerative disorders linked to angiotensin II. The 3-D structures of renin-inhibitor complexes have long been sought as an aid to the discovery of clinically effective antihypertensives.[2]

Renins[3,4] belong to the family of aspartic proteinases, several of whose structures are already known.[5-12] These structures have provided a basis for modelling mouse[13] and human renins.[14-16] More recently, X-ray analysis of an uncomplexed, partially deglycosylated form of recombinant human renin[17] has given a more accurate structure and largely confirmed the predictions of the models. However, several loop regions, which are at the periphery of the active site and may mediate interactions with the substrate, are disordered in the native human renin crystals. It is not clear whether this is a consequence of the medium resolution data, disorder in the crystal form, deglycosylation of the enzyme or the absence of a bound inhibitor.

The structures of inhibitors complexed with other aspartic proteinases[6,18-22] have defined the conformation of the extended main-chain of the inhibitor, the location and nature of the specificity subsites and the interactions of various transition state isosteres with the catalytic aspartates of these aspartic proteinases. However, a large number of questions remain concerning the specificity of renin and the differences in specificities of the human, mouse and other renins. These can be answered unequivocally only by high resolution X-ray analyses of crystals of renin-inhibitor complexes.

Although several crystal forms of inhibitor-bound mouse renin have been reported,[23-25] no high resolution analysis of a mouse renin complex has been reported. Several groups have recently obtained crystals of human renins and their complexes.[26-28] In this paper we discuss the inhibitor-complexed structures of both mouse submandibular and recombinant human renin defined in our laboratories. In parallel Grutter and coworkers have reported the crystal structures of recombinant glycosylated human renin in complex with an inhibitor.[26] These renin inhibitor structures provide a basis for understanding molecular recognition between an enzyme and substrate. They also make an important contribution towards the rational design of effective antihypertensive agents.

2 X-RAY ANALYSIS OF MOUSE AND HUMAN RENIN COMPLEXES

We have described the preparations of the two renins, the crystallization of their complexes and the X-ray data collection from the crystals elsewhere.[28] The structures were solved using the method of molecular replacement using the structure of porcine pepsin[8] as search model. Both crystal structures were refined as rigid bodies; these corresponded initially to whole enzymes, then to domains and finally to ß-strands and alpha-helices. Difference electron density maps at intermediate stages revealed the positions of the inhibitors, the saccharide moiety (human renin) and positions of loop structures unique to renins. The final agreement factors and correlation coefficients were 0.19 & 0.91 for human renin at 2.8Å resolution and 0.18 & 0.95 for mouse renin at 1.9Å resolution.[29]

For human and mouse renins the electron densities are of very high quality. The inhibitors were not included in the models used in the molecular replacement but are clearly defined in the electron density. Whereas loop structures 47-48a, 74-77, 158b-161, 241-250 and 278-282 were 'poorly ordered and had weak electron density' in the uncomplexed human renin,[17] each is clearly defined with good electron density in the inhibitor complexes of both human and mouse renins. For both human and mouse renins multiple copies of the molecules have been independently refined and have very similar structures. The coordinates of both renin structures have been deposited in the Brookhaven Data Bank.

Although human renin is N-glycosylated at two sites, mouse submaxillary renin has no N-glycosylation site and it appears that glycosylation is not essential for activity. The first site at Asn -2 in human renin has some electron density corresponding to the saccharide moiety, but this region is fully exposed to the solvent and disorder may be expected. The GlcNAc residue attached to Asn 67 is well-defined and has been built into the observed density to be refined subsequently. The density extends clearly to the second saccharide group, but becomes more diffuse beyond that.

3 OVERALL TOPOLOGY & ACTIVE SITE CLEFT

As expected from the high degree of sequence identity (~70%) of human and mouse renins, they have very similar three - dimensional structures as shown in Figure 1. Human renin models based on the structures of endothiapepsin[14] and pepsin & chymosin (Frazao & Topham, unpublished results) are seen retrospectively to have provided a reasonable description of the structure of renin. As expected, models based on pepsin and chymosin are better than those based on the more distantly related fungal aspartic proteinases.

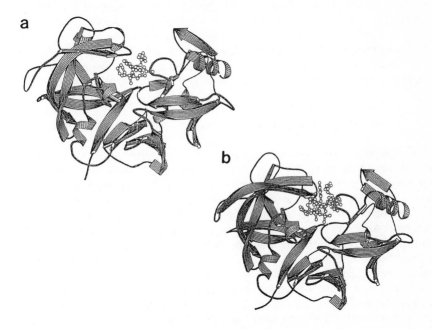

Figure 1   Schematic cartoons of (a) human and (b) mouse renin structures showing the arrangement of α-helices and ß-strands. The inhibitors are shown in full.

The active site cleft has a less open arrangement in renins than in the other aspartic proteinases. Many loops as well as the helix h_c (residues 224 - 236) belonging to the C-domain (residues 190-302)[21] are significantly closer to the active site in the renin structures compared to those of endothiapepsin-inhibitor complexes. This is partly due to a difference in relative position of the rigid body comprising the C-domain.[30] For instance, there is a rotation of ~4° and translation of ~0.1Å in the human renin complex with respect to the endothiapepsin-difluorostatone complex.

The entrance to the active site cleft is made even narrower as a consequence of differences in the positions and composition of several well-defined loops and secondary structure elements. Unique to the renins is a *cis*-proline, Pro 111, which caps the helix h_{N2} and contributes to the subsites S_3 and S_5. This helix is nearer to the active site in renins than in other aspartic proteinases. On an equivalent loop in the C-lobe (related by the intra-molecular pseudo 2-fold symmetry), there is a sequence of prolines - the Pro 292-Pro 293-Pro 294 segment. This structure is also unique to the renins among the aspartic proteinases with Pro 294 and Pro 297 in a *cis*-configuration. Such a proline-rich structure provides an effective means of constructing well defined pockets from loops which would otherwise be more flexible.

This rather rigid poly-proline loop, together with the loop comprised of residues 241-250, lie on either side of the 'flap' formed by residues 72-81. In the partly deglycosylated native renin, this 'flap' and the shorter topologically equivalent region on the C-domain, comprised of residues 241-250, were not well defined.[17] In the glycosylated, inhibitor-bound enzyme the cleft is covered by 'flaps' from both lobes rather than from the N-lobe alone as in other pepsin-like aspartic proteinases; this gives a superficial similarity to the dimeric, retroviral proteinases where each subunit provides an equivalent flap that closes down on top of the inhibitor.

4 THE ROLE OF HYDROGEN BONDS IN INHIBITOR RECOGNITION

Whereas the mouse renin inhibitor extends from P_6 to P_4', the human renin inhibitor extends only from P_4 to P_1'; the cyclohexyl norstatine residue at P_1 mimics a dipeptide analogue with its isopropyloxy group occupying the subsite for the side-chain of P_1'. Both inhibitors are bound in the extended conformation that is found in other aspartic proteinase-inhibitor complexes. Both inhibitors make extensive hydrogen bonds with the enzymes as shown in Figure 2. Although the hydrogen bonds in the two inhibitor complexes are similar from P_3 to P_1, the hydrogen bond observed between P_3 N and 219 O_γ in other aspartic proteinases is not made in the human renin complex, apparently as a consequence of a repositioning of the protonated Pro P_4 at the N-terminus of this short inhibitor. However, in general the two renin-inhibitor complexes described here demonstrate that a similar pattern of hydrogen bonding is probably used in the substrate recognition of all aspartic proteinases although their specificities differ substantially.

There is also great similarity between aspartic proteinases in recognition of the catalytic site. The catalytic aspartyl side chains and the inhibitor hydroxyl group are essentially superimposable in both complexes. The C-OH bonds lie at identical positions when the structures of inhibitor complexes of several aspartic proteinases are superposed, in spite of the differences in the sequence and secondary structure. Most of the complex array of hydrogen bonds found in endothiapepsin complexes can be formed with the exception of that to the threonine or serine at 218, which is replaced by alanine in human renin. The similarity can be extended to all other pepsin-like aspartic proteinases, and even to the retroviral proteinases, such as those from Human Immunodeficiency Virus and Rous Sarcoma Virus. This implies that the recognition of the transition state is conserved in evolution, and the mechanisms of this divergent group of proteinses must be very similar.

In the human renin complex, the outer carboxyl oxygen of Asp 215 is hydrogen bonded to the norstatine ester oxygen. We, therefore, conclude that it is protonated whilst Asp 32 is ionized and receives a hydrogen to its inner oxygen from the inhibitor hydroxyl group. This ionization state and hydrogen bond are those postulated for the stabilization of an endothiapepsin-fluoroketone hydrate complex, as well as the putative tetrahedral intermediate in proteolytic cleavage of the amide bond.[31]

There are some differences at other residues close to the catalytic aspartates that probably contribute to stabilisation of the same transition state at different pHs. The protonated and buried Asp 303 forms a hydrogen bond between its side chain carboxylic acid and the main-chain O of Thr 216. In renins this residue is alanine and as observed first by Sielecki *et al.*[17] the carboxylic hydroxyl is replaced by the less acidic phenolic hydroxyl of Tyr 155. This may still contribute to the higher pH profile of human renin. However, it appears that the higher pH profile of human renin derives also from more general changes in the electrostatic potential which may originate from the substitution of residues more distant from the catalytic residues.

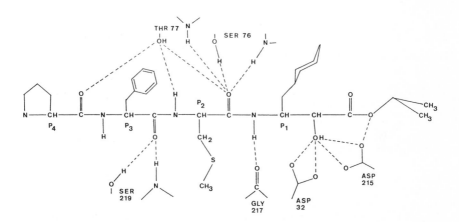

Figure 2   The hydrogen bonds between human renin and inhibitor

5 SPECIFICITY

If the main chain hydrogen bonding of substrates is conserved amongst aspartic proteinases, how are the differences in specificities achieved ? Figure 3 defines the enzyme residues that line the specificity pockets for both mouse and human renin complexes. In modelling exercises[13-16,32] it was assumed that specificities derive from differences in the sizes of the residues in the specificity pockets (S_n)[33] to complement the corresponding side chains at positions P_n in the substrate/inhibitor. A detailed analysis now shows that this simple assumption only partly accounts for the steric basis of specificity.

Subsite	Residues in contact[@] with the inhibitor	
	Human renin complex	Mouse renin complex
S6	---	S 12, F 220
S5	---	P 111, L 114
S4	T 77, L 114	S 219, F 220, H 287
S3	Q 13, T 77, P 111, L 114, A 115, F 117, A 218, S 219	Q 13, S 77, P 111, L 114, A 115, F 117, G 217, S 218, S 219
S2	S 76, T 77, G 217, S 222, M 289	Y 75, G 76, S 77, S 218, S 222, H 287, M 289, V 300
S1	D 32, Y 75, F 112, F 117, V 120, D 215, G 217	D 32, G 34, Y 75, V 120, G 217
S1'	G 34, Y 75, S 76, L 213, D 215, I 291	G 34, Y 75, G 76, V 213, D 215, V 298
S2'	---	G 34, S 35, I 73, H 74, Q 128, V 130, T 295
S3'	---	H 74, Y 75, I 291, P 292, T 295,#H 74,#R 79
S4'	---	P 294, T 295

Figure 3   Specificity pockets for the human & mouse renin complexes. [@] Distance cut-off: 4.0Å # residue from a neighbouring molecule.

For example, in the specificity subsite S_3 the phenyl rings of Phe P_3 occupy almost identical positions in both renin inhibitor complexes. Modelling studies have predicted the specificity subsite S_3 to be larger in renins than in other aspartic proteinases[32] due to substitution of smaller residues, Pro 111, Leu 114 and Ala 115, in place of larger ones in mammalian and fungal proteinases. However, a compensatory movement of the helix h_{N2} makes the pocket quite compact and complementary to the aromatic ring as shown in Figure 4a. Thus, the positions of an element of secondary structure differ between renin and other aspartic proteinases with a consequent important difference in the specificity pocket.

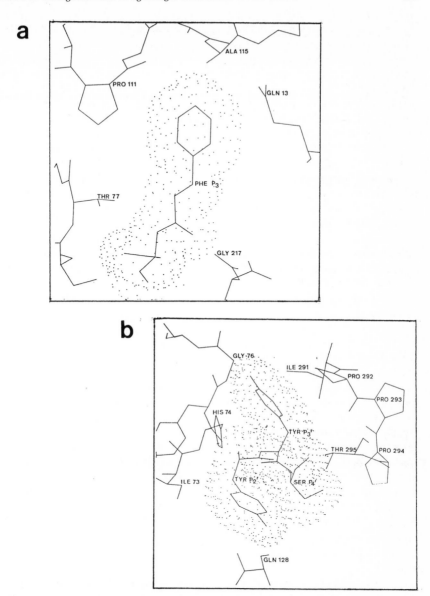

Figure 4   The specificity pockets in (a) human renin at S_3 and (b) mouse renin at S_3' and S_4'.

The differing positions of secondary structural elements may also account for the specificities at P_2'. Mouse submaxillary and other non-primate renins do not appreciably cleave human angiotensinogen or its analogues[34] which have an isoleucine at P_2', although they do cleave substrates with a valine at this position.

In contrast, human renin not only cleaves the human and non-primate substrates but also the rat angiotensinogen with tyrosine at P_2', albeit rather slowly.[35] This can be explained in terms of the three-dimensional structures. In the mouse renin complex, the P_2' tyrosyl ring is packed parallel to the adjoining helix h_3 in a narrow pocket and there is only limited space available beyond the C_8 methylene group. This appears to be able to accommodate a valine, but not the larger isoleucine at P_2' which will suffer greater steric interference from several residues which are conserved in identity and positions in the two renins. On the other hand in the human renin, differences in the orientation and position of helix h_3 bring it closer (by ~0.5Å) to the substrate-binding site than in mouse renin. It is orientated in such a fashion in human renin that, although it can accommodate the isobutyl side chain of isoleucine at P_2' , aromatic rings on substituents such as phenylalanine and tyrosine will have severe short contacts with the side chain of Ile 130 (valine in mouse renin). Thus the reorientation of a helix, coupled with subtle differences in the shapes of the side chains, makes significant changes in the substrate specificity at this subsite. It is interesting to note that in pepsin this helix is in a similar position with respect to the active site as in human renin. This provides a structural rationale for the negative influence of peptides containing phenylalanine,[36] tyrosine or histidine[37] at this subsite on proteolytic pepsin cleavage, whilst isoleucine and valine enhance catalysis.

Differences in the specificity subsites at S_1' in the human and mouse renins have a more complicated explanation. At first sight the situation appears to be explained by complementarity of the subsites to the valine and leucine at P_1' in human and mouse angiotensinogens.[13,14] Thus, residue 213 is leucine in human renin and valine in mouse renin. The S_1' pockets of chymosin, pepsin and endothiapepsin have an aromatic side-chain at residue 189 whilst the renins have amino acids with smaller side-chains (valine in human & serine in mouse renins). This would be expected to make the pocket larger in renins. However the structure of the mouse renin complex shows that the substrate moves closer to the enzyme in renins as a result of the smaller residue at 189 and the pocket is made even more compact due to a compensatory change in the position and composition of the poly-proline loop (residues 290-297). Thus, the specificity difference at this site arises not only from a compensatory movement of a secondary structure, in this case a loop region, but also from the substitution of an enzyme residue that allows the substrate to come closer to the body of the enzyme.

Elaboration of loops on the periphery of the binding cleft in renins also influences the specificity. This is most marked at P_3' and P_4', for which, it has been particularly difficult to obtain complexes with well-defined conformations for other aspartic proteinases. In endothiapepsin, which has been the subject of the greatest number of studies, different conformations are adopted at P_3' and the residue at P_4' is generally disordered. In contrast these residues are clearly defined in mouse renin. This is mainly a consequence of the poly-proline loop, illustrated in Figure 4b, which occurs uniquely in renins. The X-ray analysis of the mouse renin complex shows that the S_3'& S_4' subsites are formed by the poly-proline loop together with residues of the 'flap' and a similar situation is likely to occur in human renin. The well-defined interactions of P_3' described in the mouse renin complex explains the significant affinity when inhibitors have phenylalanine or tyrosine at P_3' as well as the importance of a P_3' residue for catalytic cleavage of a substrate by renin.[38]

Hydrogen bonds between the side chains of the inhibitor and the enzyme do not play a major role in most specificity pockets. However, S_2 is an exception. This subsite is large and contiguous with S_1' , so that in human renin the S-methyl cysteine (SMC) side-chain of P_2 is oriented towards the S_1' pocket which is only partly filled by the isopropyloxy group, the putative P_1' residue. The carbonyl oxygen of P_2 accepts a hydrogen from the O_γ of Ser 76, which is unique to human renin; residue 76 is a highly conserved glycine in all the other aspartic proteinases, including mouse renin. In mouse renin the P_2 histidyl group has a different orientation and forms a hydrogen bond with the O_γ of Ser 222. If such a conformation were adopted by the human angiotensinogen in complex with human renin, the two imidazole nitrogens would be hydrogen-bonded to the O_γ of both Ser 76 & Ser 222. This would explain why 'the catalytic efficiency and maximal velocity of human renin are greater above pH 6.5 with the substrate containing histidine at P_2 , while they are greater below pH 5.0 with glutamine[39] at this position'. The observed reduction in recombinant human renin cleavage rate of a human angiotensinogen analogue containing a 3-methyl hisitidine substituent at P_2 [40] could also be explained on the basis of the hydrogen-bonding scheme proposed above.

6 MOLECULAR RECOGNITION AND RATIONAL DRUG DESIGN

The detailed analyses of renin-inhibitor complexes reported here confirm the general structural features thought to contribute to renin's specificity[31,41] but demonstrate the need for careful, high resolution X-ray analyses for more confidence in drug design. In particular, they show that even minor alterations in the positions of secondary structural elements can lead to major changes in the disposition of the subsites and thus the recognition of substrates. Since such molecular recognition defines the species-specificity and determines the catalytic efficiency of the enzymes, a thorough understanding is indispensable for the synthesis of suitable inhibitors. The specificity pockets - the molecular recognition sites - are modified by elaboration, particularly of surface loops, which can be disordered in the uncomplexed enzymes and difficult to model with precision from homologous structures. These data establish a new foundation for the rational design of renin inhibitors in the search for a clinically successful candidate.

7 ACKNOWLEDGEMENTS

The authors thank Prof.M.Szelke, J.Sueiras-Diaz & D.M.Jones for supplying us with the mouse renin inhibitor and Prof.K.Murakami & H.Miyazaki for supplying samples of mouse renin that were used in crystallization trials. We are indebted to D.B.Damon (Pfizer Inc, Medicinal Chemistry) for synthesizing CP-85,339 and to B.W. Dominy (Pfizer Inc, Medicinal Chemistry) for discussions and modelling assistance. We thank Mr.R.Sarra (Birkbeck College) for helpful discussions. We are very grateful to Prof.N.Sakabe and A.Nakagawa for their assistance at the Photon Factory, KEK Laboratory, Tsukuba in collection of the human renin X-ray data. We thank the SERC (C.A., C.D., C.F. & M.B.), AFRC (M.N. & V.D.) and ICRF (JBC) for financial support.

8 REFERENCES

1.   R. Tigerstedt and P.G. Bergman, Scand. Arch. Physiol., 1898, 8, 223.

2.   W.J. Greenlee, Med. Res. Rev., 1990, 10, 173.

3.   P.M. Hobart, M. Fogliano, B.A. O'Connor, I.M. Schaeffer and J.M. Chirgwin, Proc. natn. Acad. Sci. U.S.A., 1984, 81, 5026.

4.   T. Imai, H. Miyazaki, S. Hirose, H. Hori, T. Hayashi, R. Kageyama, H. Ohkubo, S. Nakanishi and K. Murakami, Proc. natn. Acad. Sci. U.S.A., 1983, 80, 7405.

5.   M.N.G. James and A.R. Sielecki, J. molec. Biol., 1983, 163, 299.

6.   K. Suguna, E.A. Padlan, C.W. Smith, W.D. Carlson and D.R. Davies, Proc. natn. Acad. Sci. U.S.A., 1987, 84, 7009.

7.   T.L. Blundell, J.A. Jenkins, T. Sewell, L.H. Pearl, J.B. Cooper, S.P. Wood and B. Veerapandian, B. J. molec. Biol., 1990, 211, 919.

8.   J.B. Cooper, G. Khan, G. Taylor, I.J. Tickle and T.L. Blundell, J. molec. Biol., 1990, 214, 199.

9.   C. Abad-Zapatero, T.J. Rydel and J. Erickson, Proteins, 1990, 8, 62.

10.   G.L. Gilliland, E.L. Winborne, J. Nachman, and A. Wlodawer, Proteins, 1990, 8, 82.

11.   A.R. Sielecki, A.A. Fedorov, A. Boodhoo, N.S. Andreeva and M.N.G. James, J. molec. Biol.,1990, 214, 143.

12.   M. Newman, M. Safro, C. Frazao, G. Khan, A. Zdanov, I.J. Tickle, T.L. Blundell and N.S. Andreeva, J. molec. Biol., 1991, 221, 1295.

13.   T.L. Blundell, B.L. Sibanda and L.H. Pearl,   Nature,1983, 304, 273.

14.   B.L. Sibanda, T.L. Blundell, P.M. Hobart, M. Fogliano, J.S. Bindra, B.W. Dominy and Chirgwin, J.M. FEBS Lett., 1984, 174, 102.

15.   K. Akahane, S. Nakagawa, I. Moriguchi, S. Hirose, K. Iizuka and K. Murakami, Hypertension, 1985, 7, 3.

16.   W. Carlson, M. Karplus and E. Haber, Hypertension, 1985, 7, 13.

17.   A.R. Sielecki, K. Hayakawa, M. Fujinaga, M.E.P. Murphy, M. Fraser, A.K. Muir, C.T. Carilli, J.A. Lewicki, J.D. Baxter and M.N.G. James, Science, 1989, 241, 1346.

18.   R. Bott, E. Subramanian and D.R. Davies, Biochemistry, 1982, 21, 6956.

19.   M.N.G. James, A.R. Sielecki, F. Salituro, D.H. Rich and T. Hofmann, Proc. natn. Acad. Sci., U.S.A., 1982, 79, 6137.

20.   S.I. Foundling, J.B. Cooper, F.E. Watson, L.H. Pearl, A. Hemmings, S.P. Wood, T.L. Blundell, A. Hallett, D.M. Jones, J. Sueiras, B. Atrash and M. Szelke, J. cardiovascular Pharmacol., 1989, 10, S 59.

21.   A. Šali, B. Veerapandian, J.B. Cooper, S.I. Foundling, D.J. Hoover and T.L. Blundell, EMBO J., 1989, 8, 2179.

22.   B. Veerapandian, J.B. Cooper, A. Šali and T.L. Blundell, J. molec. Biol., 1990, 216, 1017.

23.   J.P. Mornon, E. Surcouf, J. Berthou, P. Covorl and S. Foote, J. molec. Biol., 1982, 155, 539.

24.   M.A. Navia, J.P. Springer, M. Poe, J. Boger and K. Hoogsteen, J. biol. Chem., 1984, 259, 12714.

25.   B.L. Sibanda Ph.D Thesis, 1986, University of London.

26.   J. Rahuel, J.P. Priestle and M.G. Grutter, J. Struct. Biol., 1991, 107, 227.

27.   L.W. Lim, R.A. Stegeman, N.K. Leimgruber, J.K. Gierse and S.S. Abdel-Meguid, J. molec. Biol., 1989, 210, 239.

28.   M. Badasso, B.L. Sibanda, C. Frazao, V. Dhanaraj, C. Dealwis, J.B. Cooper, S.P. Wood, T.L. Blundell, K. Murakami, H. Miyazaki, P.M. Hobart, K.F. Geoghegan, M.J. Ammirati, D.E. Lanzetti, D.E. Danley, B.A. O'Connor, D.J. Hoover, J. Suieras-Diaz, D.M. Jones and M. Szelke, J. molec. Biol., 1992, 223, 447.

29.   V. Dhanaraj, C. Dealwis, C. Frazao, M. Badasso, B.L. Sibanda, I.J. Tickle, J.B. Cooper, H.P.C. Driessen, M. Newman, C. Aguilar, S.P. Wood, T.L. Blundell, P.M. Hobart, K.F. Geoghegan, M.J. Ammirati, D.E. Danley, B.A. O'Connor and D.J. Hoover, Nature, 1992, In the press.

30.   A. Šali, B. Veerapandian, J.B. Cooper, D.S. Moss, T. Hofmann and T.L. Blundell, Proteins, 1992, 12, 158.

31.   B. Veerapandian, J.B. Cooper, A. Šali, T.L. Blundell, R.L. Rosati, B.W. Dominy, D.B. Damon and D.J. Hoover, Protein Science, 1992, 1, 322.

32.   C. Hutchins and J. Greer, Critical Reviews in Biochemistry and Molecular Biology, 1991, 26, 77.

33.   I. Schecheter and A. Berger, Biochem. Biophys. Res. Commun., 1967, 27, 157.

34.   M. Poe, J.K. Wu, Tsau-Yen Lin, K. Hoogsteen, H.G. Bull and E.E. Slater, Analyt. Biochem., 1984, 140, 459.

35.   F. Cumin, D. Le-Nguen, B. Castro, J. Menard and P. Corvol, Biochim. Biophys. Acta, 1987, 913, 10.

36.   J.C. Powers, A.D. Harley and D.V. Myers, 'Acid Proteases - Structure, Function and Biology', (ed Tang, J.), Plenum Press, New York, 1977, p. 141.

37.   V.K. Antonov, 'Acid Proteases - Structure, Function and Biology', (ed Tang, J.), Plenum Press, New York, 1977, p. 179.

38.   L.T. Skeggs, K.E. Lentz, J.R. Kahn and H. Hochstrasser, J. exp. Med., 1968, 120, 13.

39.   D.W. Green, S. Aykent, J.K. Gierse and M.E. Zupec, Biochemistry, 1990, 29, 3126.

40.   T.F. Holzman, C.C. Chung, R. Edalji, D.A. Egan, M. Martin, E.J. Gubbins, G.A. Krafft, G.T. Wang, A.M. Thomas, S.H. Rosenberg and C. Hutchins, J. Protein Chem., 1991, 10, 553.

41.   T.L. Blundell, J.B. Cooper, S.I. Foundling, D.M. Jones, B. Atrash and M. Szelke, Biochemistry, 1987, 26, 5585.

An Investigation of the Bioactive Conformation of ARG-GLY-ASP Containing Cyclic Peptides and Snake Venom Peptides Which Inhibit Human Platelet Aggregation

M. M. Hann[1], B. Carter[1], J. Kitchin[2], P. Ward[2], A. Pipe[2],
J. Broomhead[2], E. Hornby[3], M. Forster[3], C. Perry[3]

[1] COMPUTATIONAL CHEMISTRY GROUP, GLAXO GROUP RESEARCH, GREENFORD RD.,
GREENFORD, MIDDLESEX, UK
[2] MEDICINAL CHEMISTRY GROUP, GLAXO GROUP RESEARCH, GREENFORD RD.,
GREENFORD, MIDDLESEX, UK
[3] PERIPHERAL PHARMACOLOGY DEPARTMENT, GLAXO GROUP RESEARCH, PARK RD,
WARE, UK

1. INTRODUCTION

Artherosclorosis is a chronic disease of
multifactorial origin, the consequences of which may
be Thrombosis, Ischaemia, Myocardial and Cerebral
Infarction. Thrombosis arises when one or more of the
systems controlling the fluidity of blood fails,
resulting in the normal homeostasis being replaced by
clotting (coagulation). The formation of a thrombus
is critically dependent on platelet aggregation and a
key step in this process is the cross-linking of
activated platelets by the plasma protein fibrinogen.
The cross-linking of platelets is achieved by the
binding of dimeric fibrinogen molecules between
glycoprotein receptor complexes on adjacent activated
platelets. The glycoprotein receptor is one of the
Integrin family of cell adhesion molecules which are
widely utilised for this and other cell/cell and
cell/matrix interactions. The Fibrinogen Integrin
receptor is the glycoprotein GpIIB/IIIA which is
expressed on the surface of activated platelets[1]. It
is a Ca^{2+} dependent heterodimer of the glycoprotein
IIB subunits and a single chain glycoprotein IIIA
subunit. The Ca^{2+} binding domain on the α part of the
IIB subunit is believed to be the region through
which interaction with Fibrinogen takes place[2].
Fibrinogen is a dimeric protein (MW ca. 340kD) with a
blood concentration of ca. 3mg/ml in a normal
subject. It is this dimeric nature of Fibrinogen
which allows it to act as the initial bridge between
aggregating platelets.

Based on the above observations it was proposed that
inhibition of the initial binding of Fibrinogen to
its platelet receptor would lead to inhibition of
platelet aggregation and the consequential thrombus

formation. Because the Fibrinogen crosslinking is
central to platelet aggregation, inhibition of the
binding process should inhibit platelet aggregation
irrespective of the initial stimulus which activated
the platelets.

The region of the Fibrinogen molecule which is
primarily responsible for binding to its receptor has
been identified as residues 572-575, Arg-Gly-Asp (RGD
in single letter codes) from the Fibrinogen α chain[3].
The tetrapeptide RGDS inhibits platelet aggregation
in vitro with an IC_{50} of 60-80μM.

The purpose of the work described here was to explore
the conformation of RGD containing peptides which
give maximal inhibition of the binding of Fibrinogen
to its receptor - i.e. the bioactive conformation.
The ultimate aim of this work was to use this
information to help develop non-peptide, selective
Fibrinogen receptor antagonists suitable for use as
drugs for the treatment of diseases in which
excessive platelet aggregation is implicated.

2. CYCLIC PEPTIDES

METHODOLOGY

When isolated from their endogenous protein
environments, short peptides do not usually have a
defined conformation, while a defined conformation is
likely to be required for binding to the receptor.
Therefore small peptides have to loose additional
degrees of internal freedom when they bind to their
receptor as compared to the peptide in its native
protein environment. The requirement to overcome this
additional conformational freedom on binding reduces
the amount of free energy available on binding; the
"cost" of this has been estimated as 0.7 kcal/mol[4]
per bond able to undergo free rotation in the
uncomplexed ligand, but fixed in the complex. If the
bioactive conformation can be preinduced into a small
peptide then an increased binding affinity to the
receptor, equivalent to this "cost", should be
observed. The corollary of this is that such an
increase in binding affinity among a series of
analogues is indication that the preferred
conformation has been induced in the ligand prior
to binding. Peter Kollman referred to this in his
talk earlier this week as "paying the price in
synthesis".

Secondary structure analysis of the RGD region in
several Integrin receptor ligands (including
Fibrinogen) indicated that the RGD motif may be

involved in a surface oriented β-turn. Because
Glycine is often observed at the corner of a β-turn,
we postulated that if we could preinduce such a
conformation into small RGD peptides then an increase
in binding affinity would be observed. The literature
contains numerous examples of β-turn mimics but
rather than try and build the RGD sequence onto one
of these mimics a different approach was adopted.
This was because a large amount of synthetic effort
would be required to make such compounds, with the
Arginine or Aspartic acid side chain functionalities
built onto the actual turn mimics. We were interested
in testing our hypothesis by the most pragmatic
approach and this would inevitably mean making a
number of analogues. The alternative approach that we
adopted was to put the RGD sequence into cyclic
penta- and hexa-peptides and to use the nature of the
additional residues necessary to cyclise the RGD tri-
peptide to induce different conformations into the
RGD region. In this way the additional residues can
be considered as molecular spanners which adjust the
conformation of the RGD part. Based on the above
strategy a series of cyclic peptides was synthesised
of which compounds 2-7 are representative examples.
The structures and fibrinogen antagonist activity of
these compounds are listed in Table 1.

Table 1. Cyclic RGD peptides and their Fibrinogen
         antagonist activity

Compound	Sequence*	Activity+
(1)	GRGDS	1.0 (standard)
(2)	<RBDFG>	>44
(3)	<RGD-(Me)FG>	>25
(4)	<RGDFG>	0.4
(5)	<RGDF-[D]PF>	0.7
(6)	<RGDF-[c65]>	0.9
(7)	<RGDF-[t65]>	0.02

Key: < > = cyclic peptide
     *   standard single letter amino acid codes
         with the following additions:
         B is β- Alanine.
         (Me)F is N-Methyl-Phenylalanine
         [D]P is D-Proline.
         [c65] & [t65] = dipetide mimics (see below)
     + Activity is expressed relative to compound
        (1) which has an IC_{50} of 53μM against ADP
        induced aggregation of gel filtered human
        platelets.

The simplest of the cyclic compounds contains the RGD
motif in a cyclic pentapeptide (cyclo-Arg-Gly-Asp-
Phe-Gly (4)). The related cyclic pentapeptide (cyclo-
Arg-β-Ala-Asp-Phe-Gly (2)) has a β-Alanine residue
replacing the Glycine residue at position 2. (nb
sequences are numbered with Arg at position 1). In
cyclic pentapeptide (3), (cyclo-Arg-Gly-Asp-(N-
Me)Phe-Gly), the Phenylalanine nitrogen atom is N-
methylated. The cyclic hexapeptide (5), (cyclo-Arg-
Gly-Asp-Phe-D-Pro-Phe) has a D-Phenylalanine residue
at position 5. Compounds (6) and (7) are cyclic
hexapeptides in which the 5th and 6th residues are
fused together into a bicyclic system in which a δ-
lactam is fused to a thiazolidine ring as shown
below. Compounds (6) and (7) are diastereoisomeric -
compound (6) has a <u>cis</u> relationship between the
substituents to the six-five ring system (referred to
as [c65]) while compound (7) contains the [t65]
fragment in which the substituents are <u>trans.</u>

(6)   (7)

The synthesis of the 6,5 bicyclic fragments ([c65]
and [t65]), and their incorporation into (6) and (7)
is outlined in schemes I and II. The <u>cis</u> bicyclic
fragment [c65] is the expected product of this
reaction scheme based on the L-amino acids used but
ca. 15% of the <u>trans</u> isomer [t65] is produced by
racemisation at the stage of the ester hydrolysis.
Previous reported syntheses of the [c65] fragment[5]
have only yielded the <u>cis</u> isomer but the racemisation
observed in the synthesis that we used, and which
gave the [t65] compound (7), turned out to be highly
fortuitous (see below). After Fmoc protection, the
bicyclic fragments and the other amino acids were
assembled into the acyclic precursors by standard
solid phase peptide synthesis technology.

Scheme I Synthesis of bicyclic fragments [c65] and [t65]

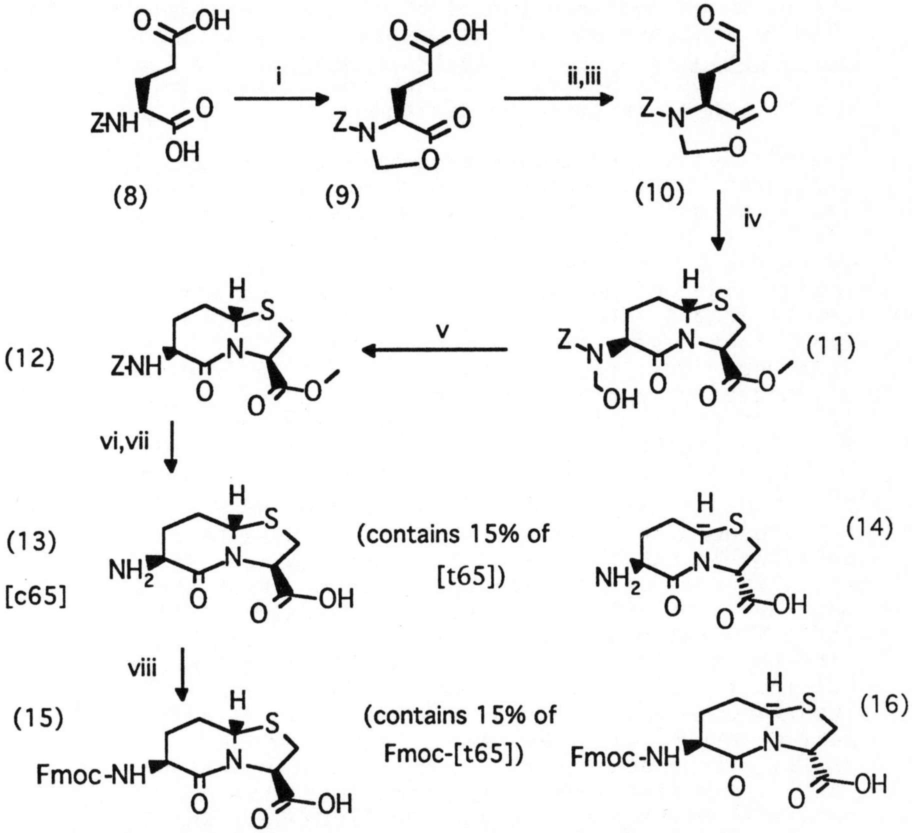

Reagents: i Tosic acid, $(CH_2O)_n$, Toluene. ii $ClCH=N^+Me_2Cl^-$, Pyridine, THF
iii $Li(Bu^tO)_3AlH$, THF, CuI. iv L-CysOMe.HCl, Pyridine. v Na_2CO_3, MeOH.
vi NaOH, MeOH, H_2O. vii HBr, AcOH. viii Fmoc.Cl, dioxan, Na_2CO_3 (aq).

DETERMINATION OF THE SOLUTION CONFORMATION OF THE CYCLIC PEPTIDES

The solution conformation of the cyclic peptides in DMSO or DMSO/sulpholane was determined by using nmr and computational chemistry techniques. 2D COSY and NOESY or ROESY spectra were used for assignments and derivation of interproton distances.

The nOe cross peak intensities were converted to distances by reference to the observed cross peak intensities for protons of known distances apart (eg Glycine or Proline methylene groups) using a $1/r^6$ relationship. A database of possible cyclic hexaglycine and pentaglycine conformations was constructed in Chem-X as Foreign Conformational Analysis files[7] using the data of Paul et al.[8]. The observed inter-proton distances (+/- 0.7Å) for a given cyclic peptide were then used to search the database to find structures which would act as suitable starting templates for further refinement. Using Chem-X the hexaglycine peptide template was then modified such that a β carbon atom was reintroduced at positions where there were D- or L-residues in the peptide under consideration (Proline residues were also explicitly created). With the peptides containing the bicyclic restraints [c65] and [t65] the relevant structural fragments were built into the cyclic structures by use of the molecular editing facilities within Chem-X. The Chem-X interface to the molecular mechanics and dynamics package AMBER[8] was then used to set up AMBER minimisations and restrained dynamics simulations. Initial minimisations, with harmonic restraints proportional to the observed nOe distances, were carried out on the conformations selected from the cyclic peptide database (the restraints force constant was $10kcal/A^2$). This was done to remove initial high strain values which would make the restrained dynamics simulations unstable. After minimisation for 100 cycles of steepest descent, a restrained dynamics simulation was carried out for 200ps at a temperature of 700°K with the structures being written to disk every 5ps. After the dynamics was complete the simulation was viewed within Chem-X to ensure that nothing untoward had happened during the simulation. The saved structures from the simulation were all then subjected to energy minimisation, again with the nOe distances used as harmonic restraints. The conformation which had the lowest internal energy and the lowest restraints energy was considered to be the most representative of the solution conformation of the peptide. This

conformation was then used as the basis for further
addition of the side chain functionality.

Checks on calculated (using Macromodel[9]) and observed
coupling constants and further distances (eg backbone
to side chain) not used as constraints in the
calculations were also used to evaluate and validate
the selected conformations. Additionally the
temperature dependence of the amide NH resonances was
used as an indication as to whether an NH proton was
involved in a hydrogen bond and this was checked for
consistency with the calculated conformations. While
there was usually a good correspondence between the
observed and calculated data, it is likely that the
"snap shot" conformations selected and reported here
are representative of a family of conformations.
Indeed viewing the molecular dynamics simulations
showed that, while more conformationally restrained
than linear peptides, these cyclic peptides retain
some flexibility. Table 2 shows a descriptive summary
of the conformations found for the structures and Fig
1-6 show stereo representations of the conformations.

Table 2. <u>Solution conformations of RGD containing
         peptides</u>

Compound	Sequence	Activity[1]	Turn Position[2]
(2)	<RBDFG>	I	βII at <u>DF</u>,
(3)	<RGD-(Me)FG>	I	γ at <u>(Me)F</u>, <u>G2</u> and <u>R</u>
(4)	<RGDFG>	A	γ at <u>G2</u> and <u>G5</u>
(5)	<RGDF-[D]PF>	A	βII' at <u>GD</u>, γ at <u>E</u>
(6)	<RGDF-[c65]>	A	βII' at <u>GD</u>, βII' at [c<u>651</u>, γ at <u>E</u>
(7)	<RGDF-[t65]>	M	βI at <u>DF,</u> γ at [t6<u>5</u>] γat F, βII' at [t6<u>51R</u>

Notes:
1. Activity: I=Inactive, A=Active, M=More active
2. eg.γ at <u>G2</u> means γ-turn at Gly-2 (numbering
      always starts at Arg-1)

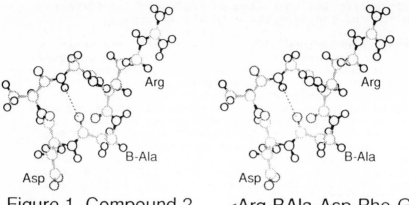

Figure 1. Compound 2   <Arg-BAla-Asp-Phe-Gly>

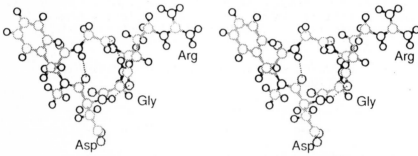

Figure 2. Compound 3   <Arg-Gly-Asp-(NMe)Phe-Gly>

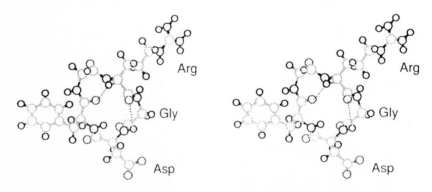

Figure 3. Compound 4   <Arg-Gly-Asp-Phe-Gly>

DISCUSSION

In Table 2 the Fibrinogen antagonist activities have been grouped into 3 classes - inactive (I), active (A) and more active (M). This classification derives from the data shown in Table 1 where it can be seen that compounds 2 and 3 are inactive while compounds 4, 5 and 6 have activities similar to the standard GRGDS acyclic peptide. Compound 7 is described as more active because it is approximately two orders of magnitude more potent than GRGDS. The three active compounds all have a βII' or γ turn at the Gly-Asp residues and when overlayed these three compounds all position the Asp and Arg side chains in similar regions of space. Glycine commonly occurs at the corner of a β-turn and the original hypothesis of this work was that such a turn was likely to be the bioactive conformation. However the compounds with Gly at the corners of a turn are only equiactive with the GRGDS standard and, as their conformational flexibility is greatly reduced compared to the acyclic GRGDS, it is unlikely that the conformations suggested for these 3 compounds is representative of the bioactive one.

The two inactive compounds, (2) and (3) have a similar overall shape to the three compounds discussed above. However, while compounds (4), (5) and (6) have the carbonyl group of the amide between Asp and Phe on the rear face of the molecule (as depicted in Figures 4, 5 & 6, the two inactive compounds have the carbonyl group on the front face. This is the most obvious distinguishing characteristic of these two compounds and the accompanying reversal of the dipole in this region of the molecule could be sufficient to completely switch off binding to the receptor.

Compound (7), which is derived from the fortuitous racemisation product [t65], is approximately two orders of magnitude more potent than the compounds (4), (5) and (6) and the solution conformation deduced for this compound is clearly different to that of the other compounds. It is characterised by an extended conformation along the Arg-Gly-Asp side of the cyclic peptide which helps to hold the Arg and Asp side chains as far apart as possible. A βII' turn holds the Asp-Phe amide bond in a similar orientation to that of compounds (4), (5) and (6) - that is with the carbonyl on the back face. The temperature shifts of the amide protons in (7) suggests that the Gly and [t65] protons are involved in intramolecular hydrogen

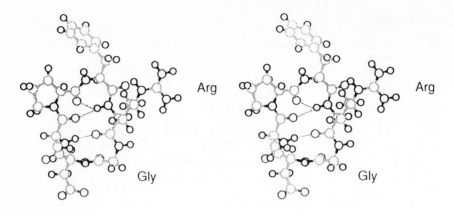

Figure 4. Compound 5 <Arg-Gly-Asp-Phe-DPro-Phe>

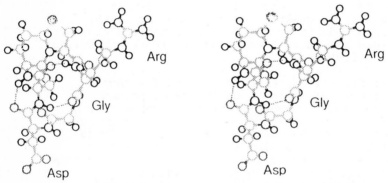

Figure 5. Compound 6   <Arg-Gly-Asp-Phe-[c65]>

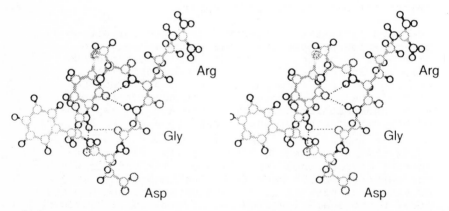

Figure 6. Compound 7   <Arg-Gly-Asp-Phe-[t65]>

bonds and this is consistent with the positions of the two β-turns in the conformation deduced.

Figure 7 shows a superimposition of compounds (7), (6) and (2) as representative examples of each class of compounds - I , A and M. This superimposition was done by an RMS fit of the equivalent backbone atoms of the Gly and Asp residues. The "incorrect" inverted orientation of the Asp-Phe amide bond for the inactive compounds is clearly seen when compared to that of the active compounds. The extended conformation of the RGD portion of the more active compound (7) is seen to be as a result of the [t65] moiety adopting the i and i+1 postions of the βII' turn. This is in contrast to the [c65] moiety in compound (6) which occupies the i+1 and i+2 of the βII' turn, resulting in a shift of the sequence against the underlying turns of the cyclic hexapeptide.

The 2 fold increase in activity of (7) over the linear analogue (1) is equivalent to an additional 2.8 kcal/mol of binding energy available to (7). If this is due to the correct conformation having been induced in (7) then 2.8 kcal/mol is equivalent to 4 bonds being frozen into the correct conformation (at ca. 0.7 kcal/mol/bond). This is consistent with a linear bioactive conformation for Arg-Gly-Asp in which the Arg Ψ, Gly Φ and Ψ, and Asp Φ torsional angles have been correctly set.

Also included in Fig 7 is a representation of the relationship of the Arg to Asp binding distance. This is ca.14A, for the conformation of the side chains shown. The Arg binding site is represented as an ion pair to a carboxylate group of the IIB/IIA complex and the Asp binding site as the missing coordination site of a putative Calcium binding site on the receptor complex (as suggested by Poncz et al.[2]).

3 ECHISTATIN

The other approach that we took to try and find out relevant information about the bioactive conformation of RGD peptides when bound to Integrin receptors was to explore the conformation of Echistatin, a representative example of the snake venom family of disintegrin molecules[10]. Echistatin has potent (IC50 ca 1nM) inhibitory activity at the GpIIb/IIIa receptor and we believed that this was likely to be due to the RGD moiety being held in a preferred conformation for binding - presumably similar to that which we deduced from the cyclic peptide work.

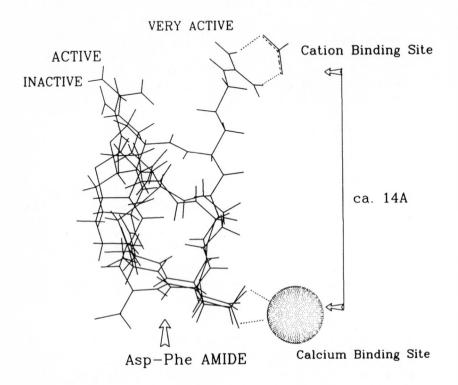

Figure 7. Overlay of different classes of compounds

Using 2D nmr techniques and distance geometry and simulated annealing techniques, members of our Protein Structure Group and collaborators at the University of York derived a solution structure for Echistatin[11]. To our surprise, while the core of the molecule was well defined, the part of the sequence containing the RGD motif did not have a well defined conformation. Figure 8 shows an RMS superimpostion of several solutions arising from the computational techniques used. The loop containing the RGD motif is visible at the top of the structure and it can be clearly seen that this is where there is considerable variation in the conformation of the different solutions, which we interpret as conformational flexibility. How is it then that Echistatin and the other disintegrins appear to be so potent as Fibrinogen antagonists? There are several possible explanations, but the two most plausible are that there are other residues in Echistatin (apart from the RGD motif) which contribute to binding or that, despite the apparent flexibility of the RGD region in Echistatin (as deduced from the nmr experiments), there are in fact a limited number of conformations available which the Integrin receptor is able to select from. This latter suggestion is consistent with Echistatin being a potent inhibitor at a number of different Integrin receptors which may require slightly different conformations. This also ties in with our observations on the selectivity of compound (7) for the Fibrinogen receptor and the recent observations of Kessler et al. that cyclic peptides with turns in the middle of the RGD portion are selective for the Vitronectin receptor[12].

An interesting recent observation brings a further twist to the Echistatin story. Blobel et al.[13] have shown that a domain of the protein PH-20, which occurs on the surface of mammalian sperm, is homologous to the snake venom Disintegrins and it is thus likely to have a similar 3D fold to that which we have deduced for Echistatin. However the RGD motif of the Disintegrin domain in PH-20 is replaced by a TDE(CD) motif. This alternative motif is suggested as the ligand by which the sperm attaches to an Integrin type receptor on an egg. The story of Adam and Eve in the Garden of Eden should perhaps be reavaluated in the light of this!

4 CONCLUSION

We have explored the preferred bioactive conformation of the RGD motif for binding to the GpIIb/IIIa receptor. These experiments in molecular recognition have been done without any knowledge of the structure

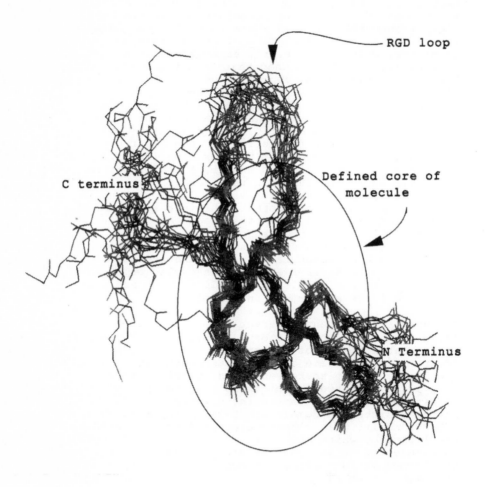

Figure 8. View of the backbone atoms of 24 Echistatin structures derived from simulated annealing of the NMR data

of either of the partners in the interaction. Despite this we are able to conclude that an extended conformation between the Arg and Asp is the bioactive conformation and we are currently utilising this information in our continuing program to develop non-peptide Fibrinogen antagonists.

ACKNOWLEDGEMENTS

Thanks are due to Drs P.Murray-Rust and R.Cooke (Protein Structure Group, GGR) and Drs P.Herzyk, M Hartshorn and R.Hubbard (University of York) for the Echistatin structure and Drs A.Tonge and R.Newton for helpful discussions.

REFERENCES

1. E.F.Plow, M.H.Ginsberg and G.A.Marguerie in 'Biochemistry of Platelets', D.R.Phillips, M.R.Shuman, Academic Press, New York, 1986.
2. M.Poncz, R.Eisman, R.Heidenreich, S.M.Silver, G.Vilaire, S.Surrey, E. Scwartz and J.S.Bennett, J.Biol.Chem., 1987, 262, 8476.
3. M.D.Pierschbacher and E.Ruoslahti, Nature, 1984, 309, 30.
4. P.R.Andrews, D.J.Craik, J.L.Martin, J.Med.Chem, 1984, 27, 1648.
5. U.Nagai and K.Sato, Tet.Lett., 1985, 26, 647
6. Chem-X molecular modelling software, Chemical Design Ltd., Unit 7, Westway, Oxford.
7. P.K.C.Paul and C.Ramakrishnan, Pep.Prot.Res., 1987, 433.
8. S.C.Weiner, P.A.Kollman, D.T.Nguyen and D.A.Case, J.Comp.Chem., 1986, 7, 230.
9. F.Mohamadi, N.J.Richards, W.C.Guida, R.Liskamp, M.Lipton, C.Caufield, G.Chang, T.Hendrickson, W.C.Still, J.Comp.Chem., 1990, 11, 440.
10. V.M.Garsky, P.K.Lumma, R.M.Friedinger, S.M.Pitzenberger, W.C.Randall, D.F.Weber, R.J.Gould and P.A.Friedman, Proc.Nat.Acad.Sci.USA, 1989, 96, 4022.
11. R.M.Cooke, B.G.Carter, D.A.Martin, P.Murray-Rust and M.Weir, Eur.J.Biochem, 1991, 202, 323.
12. M.Aumailley, M.Gurrath, G.Muller, J.Calcete, R.Timp and H.Kessler, FEBS, 1991, 291, 50.
13. C.P.Blobel, T.G.Wolfsberg, C.W.Tuck, D.G.Myles, P.Primakoff and J.M.White, Nature, 1992, 356, 248.

Bacterial IgG-Binding Proteins: Two Solutions to a Molecular Recognition Problem

G. C. K. Roberts, L.-Y. Lian, J. C. Yang, J. P. Derrick and
M. J. Sutcliffe

BIOLOGICAL NMR CENTRE AND DEPARTMENT OF BIOCHEMISTRY, UNIVERSITY OF
LEICESTER, LEICESTER LE1 9HN, UK

1. INTRODUCTION

A number of species of pathogenic bacteria, notably *Streptococci* and *Staphylococci*, have proteins on their surface which are able to bind to the constant region (F_c) of immunoglobulin G (IgG).[1,2] These proteins have attracted considerable interest, both because of their role in enhancing microbial virulence and, at the practical level, because of their use in a wide variety of immunochemical methods. The best known is protein A from *Staphylococcus aureus;* this contains five highly homologous regions, each 58 residues in length, which are able to bind specifically to the F_c (rather than the F_{ab}) portion of IgG. Detailed structural information on this interaction has come from the solution of the crystal structure of a complex between a single IgG-binding domain of protein A and a human F_c fragment.[3] In this structure, two α-helices of the protein A domain make extensive contacts with residues at the interface between the C_H2 and C_H3 domains of F_c. A recent analysis of the secondary structure of one of the IgG-binding domains of protein A in solution[4] provided evidence for the existence of a third α-helix in the molecule, which was postulated to be disrupted on formation of the complex with F_c.

Protein G is an IgG-binding protein associated with their cell walls of species of *Streptococcus*; this has a high affinity for immunoglobulins from a wide range of mammalian sources.[5] Complete or partial DNA sequences for the protein G genes from several *Streptococci* have been reported;[6-9] the molecule is approximately 600 residues long, and contains regions responsible for binding serum albumin as well as those responsible for binding immunoglobulin G. As in the case of protein A, the antibody-binding regions of protein G have been identified as two or three (depending on the *Streptococcal* strain examined) closely homologous domains. Expression of a truncated form of protein G, containing only the three IgG-binding domains,[9] or of single IgG-binding domains,[6,7] in *E. coli* has been shown to generate a protein which retains a high affinity for IgG. The IgG-binding domains of protein G are 55 residues long, and separated from one another by short, 15-residue, "linker" sequences[6,10] (in the latter respect the arrangement of the domains differs from that in protein A). There is no discernible sequence homology between the IgG-binding domains of protein G and those of protein A. Strikingly, the binding of protein G to F_c is competitive with that of protein A, suggesting that the two proteins bind to the same site on F_c.[11] This raises the possibility that the two kinds of domain may be structurally similar in spite of the lack of similarity in their sequences; we have therefore used nmr spectroscopy to determine the three-dimensional structures of the IgG-binding domains of protein G in solution. Specifically, we have studied protein G form the *Streptococcus* strain GX7805, which has three IgG-binding domains, the first two identical in aminoacid sequence, while the third differs by just four aminoacid substitutions.

2. STRUCTURE DETERMINATION

The stretches of DNA coding for domains II and III of protein G, designed to include parts of the "linker" sequences on either side of the domains, were separately replicated from pSPG29[9] and inserted into the expression vector pUC18; levels of 50-100 mg protein were obtained from 1 l cultures, and were purified as described.[12]

NMR spectroscopy

The procedures involved in the determination of the three-dimensional structure of a small protein by nmr are summarised in Figure 1; the first essential step is to assign all the resonances in the ^1H nmr spectrum. In the case of domains II and III this was done[12,13] by using a combination of homonuclear two-dimensional TOCSY and NOESY spectra, following essentially the sequential assignment strategy of Wüthrich.[14] Briefly, this strategy consists of two steps: First, the type of amino-acid residue from which a particular resonance arises is identified, by analysis of patterns of cross-peaks due to scalar coupling (a *through-bond* connection) in two-dimensional COSY or TOCSY spectra. In COSY spectra, connections are observed between the resonances of protons separated by two or three bonds, while in TOCSY spectra connections can often be observed along the whole length of an amino-acid sidechain. In the second step, sequence specific assignments within each amino-acid type ('sequential assignments') are made by observation of NOEs (a *through-space* effect) between the NH proton of one residue and protons of the preceding residue in the sequence. Typically, NOEs are seen between protons within 5Å of one another. The local conformation of the polypeptide chain determines which proton of residue i will be within <5Å of the NH of residue $i+1$, but whatever the conformation, an NOE is expected to at least one of the NH, CαH or CβH protons. Figure 2 shows part of a NOESY spectrum of the protein G domain in which a series of NH_i-NH_{i+1}

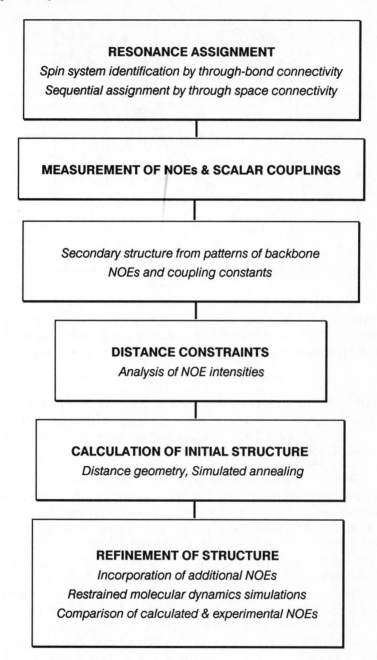

Figure 1. Steps in structure determination by nmr

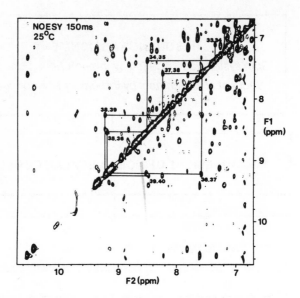

Figure 2. NH-NH region of the NOESY spectrum of domain I of protein G in 90% H_2O/10% ^2H_2O, pH 4.2, 25°C, recorded with a mixing time of 150 ms. The NH-NH sequential connectivities for residues 33-40 are indicated. From [13]

NOEs allow one to trace connections along a section of the protein backbone. Since the type of amino-acid residue to which each of these NH protons belong has been identified in the first stage of assignment, one can position this section within the sequence of the protein - in this case, residues 33-40.

Because the distances between protons of adjacent residues in the sequence depend on local conformation, the sequential NOEs used to make resonance assignments can also provide information on secondary structure. For example, a sequence of strong NH_i-NH_{i+1} NOEs, such as that seen in Figure 1, and $C\alpha H_i$-NH_{i+3} NOEs, accompanied by weak $C\alpha H_i$-NH_{i+1} NOEs, is characteristic of a helix, while strong $C\alpha H_i$-NH_{i+1} NOEs are characteristic of an extended conformation such as seen in a β-sheet. One can thus obtain a picture of the secondary structure of the protein from a summary of the sequential NOEs such as that presented in Figure 2. The $C\alpha H_i$-NH_{i+1} NOEs indicate the existence of an extended conformation from residues 4 to 26, interrupted by a turn at residues 15-16; there is then a helix from 28 to 43, characterised by strong NH_i-NH_{i+1} NOEs, followed by extended conformation for the remainder of the chain except for a loop around residue 49. Longer-range NOEs - between residues which are not adjacent in the sequence - indicate that the four stretches of extended conformation identified in this way do indeed form a four-stranded β-sheet;[13] this already indicates a difference in structure from the IgG-binding

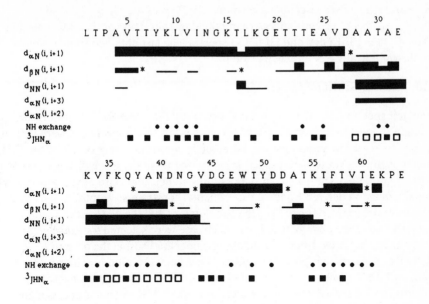

Figure 3. Summary of short-range NOEs involving NH, CαH and CβH protons of domain I of protein G. The relative strengths of the NOEs are indicated by the thickness of the bars; potential connectivities which were obscured by overlap are indicated by an asterisk. The Figure also includes information on the magnitude of some of the $^3J_{HN-C\alpha H}$ coupling constants (filled square: >8Hz, open square: < 6Hz) and on the exchange of amide protons (filled circles indicate amide protons which did not exchange after incubation in ^2H_2O for 12 h at 25°C). From [13]

domain of protein A, which contains no β-sheet structure.

To go further and construct a model of the three-dimensional structure of the protein in solution, approximate information on internuclear distances is obtained by semi-quantitative analysis of the experimental NOEs. Typically, strong. medium and weak NOEs are taken as corresponding to maximum internuclear distances of 2.5, 3.5 and 5Å. In the present case, the cross-peaks in the NOESY spectra were well-resolved, and a more detailed analysis of the intensities in NOESY spectra obtained at different mixing times was carried out. Upper bounds for the distance restraints were thus classified on the basis of NOE intensity as < 2.4Å, < 2.9Å, < 3.4Å, < 4.0Å and < 5.5Å; all lower distance bounds were set to 1.8Å.

Additional conformational information in the form of constraints on dihedral angles, particularly about the Cα-N and Cα-Cβ bonds, can be obtained from measurements of scalar coupling constants. Values of the $^3J_{HN-C\alpha H}$ and $^3J_{C\alpha H-C\beta H}$ coupling constants were obtained by simulation of the cross-peaks in DQF-COSY, PE-COSY and HMQC-J spectra (the last obtained from a sample of biosynthetically ^{15}N-enriched protein).[12]

Calculation of Structural Models

The nmr spectra yielded, as outlined above, a total of 478 NOE constraints for domain II and 445 for domain III, of which 151 and 142, respectively, were long range ($i\rightarrow>i+4$); in addition, 40 dihedral angle constraints were obtained for each domain.[12] These were used as input for the torsional space distance geometry program DIANA[15], which produces three-dimensional structures satisfying a penalty function derived from the experimental constraints. A total of 100 structures were determined for each of the two domains and the 40 structures with the lowest value of the penalty function in each case were selected for subsequent refinement, involving dynamical simulated annealing[16] using the program XPLOR; the particular procedure adopted has been reported.[12] This procedure of using DIANA followed by simulated annealing was repeated in an iterative manner until there were no large violations of the NMR restraints and as many as possible of the ambiguous assignments of NOESY cross-peaks had been resolved on the basis of proton-proton distances observed across the ensemble of generated structures. The final stage in the protocol was the use of restrained molecular dynamics refinement.[12] During this stage (which again used XPLOR) the NMR derived restraints were combined with a full potential *in vacuo* molecular dynamics force field. The "final" structure in each case was obtained by averaging the conformations over the later part of the trajectory at 300K and energy minimising this average structure. The final refined structures derived from each of the 40 DIANA structures for each domain were superimposed by least squares; the root mean square differences between these "families" of structures provide a rough measure of the precision of the structure determination.

3. STRUCTURES OF BACTERIAL IgG-BINDING PROTEIN DOMAINS

The structure obtained[12] from nmr for domain II of protein G is shown in Figure 4, where it is compared with the structure of a domain of protein as seen in its complex with F_c.[3]

The structure of domain III of protein G is essentially identical to that of domain II. In both cases, the N- and C-terminal regions (residues 1-5 and 61-64), which correspond to parts of the "linker" sequences, are poorly defined, and the loop regions are less well defined than the secondary structural elements. Omitting the poorly defined N- and C-terminal regions, the root mean square deviation of the backbone atoms (N, Cα and CO) with respect to the average structure is 0.8 ± 0.3Å for domain II and 0.9 ± 0.3Å for domain III, and that of all

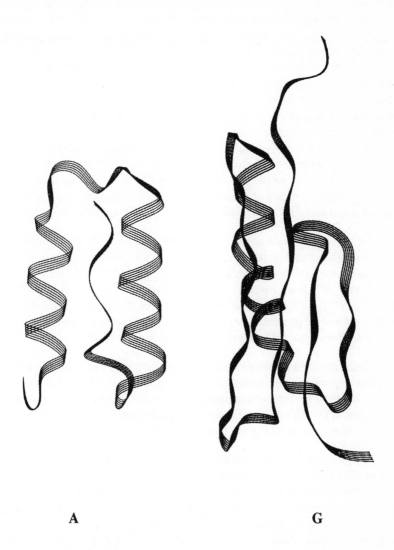

Figure 4. A comparison of the structures of the IgG-binding domains of protein A (left, from original structure in[3]) and protein G (right, domain II, from[12]). The structures are shown as backbone ribbon traces to emphasise the secondary structure; in the case of the protein G domain, the structure shown is the average of the 40 structures calculated using the experimental constraints.

the heavy atoms is 1.7 ±0.3Å for domain II and 1.9±0.3Å for domain III. It is clear that the two double stranded antiparallel β-sheets evident from the initial NOE analysis (Figure 3) interact in a parallel manner to form an unusual four stranded antiparallel-parallel-antiparallel β-sheet, with the N- and C-terminal strands in the centre of the sheet. The helix, residues 28-40, runs essentially diagonally across the sheet, packing against it to form a solvent inaccessible core comprising the hydrophobic residues Leu 10, Phe 35, Trp 48, Phe 57 and Val 59. Although there is no direct evidence to allow differentiation between an α−helix and a 3_{10}-helix, the indirect evidence from the hydrogen bonding pattern indicates that the helix is likely to be entirely α−helical in nature. Among the proteins of known structure, the folding topology of these domains of protein G is most similar to that of ubiquitin,[17] which has a similar antiparallel-parallel-antiparallel β-sheet with a helix lying diagonally across it, but which has a longer loop and a short fifth strand of β-sheet inserted between strands three and four. Even within the elements of secondary structure which correspond between ubiquitin and protein G, however, there is only 12% sequence identity.

The topology of both domains studied here is similar to that reported[18] for a shorter IgG-binding domain from protein G from *Streptococcus* strain GX7809, which is closely similar in sequence to residues 6-61 of the domain II studied here, having only two conservative amino acid substitutions in the first β-strand (V11I and I12L). Despite these small differences in sequence, the structures of domains II and III presented here do differ in some respects from that presented by Gronenborn *et al.*;[18] these are discussed in detail elsewhere.[12]

The comparison in Figure 4 shows very clearly that the structures of the domains of protein A and protein G, which appear to bind to the same site on F_c, have quite different three-dimensional structures, as well as different amino-acid sequences.* This therefore represents an interesting example of *two structural solutions to the recognition of the same region of a protein surface.*

The only real point of similarity in structure between the protein A and protein G domains is in the presence of a helix which, in the case of protein A, is involved in binding to F_c. This suggests the possibility that the helix of protein G might similarly be involved in binding. Some slight support for this notion comes from a comparison of protein G domains II and III, which differ in their specificity for binding to different classes of IgG.[20] As noted above, these two domains have essentially the same conformation, differing slightly at only one point, in the loop Gly 43-Asp 45. It is therefore likely that the difference in specificity arises because one or more of the residues which differ between the two domains is involved directly in the binding process. These differences are as follows: E24K, at the end of the second strand of β-sheet, A29E and V34A in the α-helix, and E47V, at the beginning of the third strand of β-sheet, but on the

---

* The comparison with ubiquitin mentioned above illustrates that similar folding patterns *can* be found in proteins of unrelated sequence.[19]

other side of the structure from residues 24, 29 and 34. The fact that two of the four differences are in residues in the helix is at least consistent with the idea that the helix of the protein G domain is involved in binding to F_c.

From the three different protein G domains for which structures are now available, it is apparent that several substitutions of surface residues can be made without altering the overall structure of the protein. This opens the way to the "engineering" of the molecule. We are currently examining the individual contributions of each of the surface residues in the helix to F_c binding by site-directed mutagenesis. This will permit identification of residues which interact with F_c and, we anticipate, will also allow the design of protein G domains with novel specificities. We are also studying the protein G domain/F_c binary complex using stable isotope-edited nmr techniques; this affords a more direct approach to the identification of the residues involved and to developing a model for this protein-protein interaction.

4. ACKNOWLEDGEMENTS

This work was supported by a Wellcome Trust Fellowship to JPD, and used the facilities of the Leicester Biological NMR Centre, which is supported by SERC. We are grateful to Drs. C.R. Goward, J.P. Murphy and T. Atkinson (PHLS, Porton Down) for the cDNA encoding the three IgG-binding domains of protein G from *Streptococcus* strain GX7805.

5. REFERENCES

1. J. J. Langone, *Adv. Immunol.*, 1982, *32*, 158.
2. M. D. P. Boyle, ed., 'Bacterial Immunoglobulin Binding Proteins', Academic Press, New York, 1990.
3. J. Deisenhofer, *Biochemistry*, 1981, *20*, 2361.
4. H. Torigoe, I. Shimada, A. Saito, M. Sato and Y. Arata, *Biochemistry*, 1990, *29*, 8787.
5. L. Björck and B. Åkerström, in 'Bacterial Immunoglobulin Binding Proteins', (M.D.P. Boyle, ed.) Academic Press, New York, 1990, p. 113.
6. S. R. Fahnestock, P. Alexander, D. Filpula and J. Nagle in 'Bacterial Immunoglobulin-Binding Proteins' (M.D.P. Boyle, ed.), Academic Press, New York, 1990, p. 133.
7. B. Guss, M. Eliasson, A. Olsson, M. Uhlén, A.-K. Frej, H. Jörnvall, J.-I. Flock and M. Lindberg, *EMBO J.*, 1986, *5*, 1567.
8. A. Olsson, M. Eliasson, B. Guss, B. Nilsson, U. Hellman, M. Lindberg and M. Uhlén, *Eur. J. Biochem.*, 1987, *168*, 319.
9. C. R. Goward, J. P. Murphy, T. Atkinson and D. A. Barstow, *Biochem. J.* 1990, *267*, 171.
10. U. Sjöbring, L. Björck and W. Kastern *J. Biol. Chem.*, 1991, *266*, 399.

11. G. C. Stone, U. Sjöbring, L. Björck, J. Sjöquist, C. V. Barber and F. A. Nardella, *J. Immunol.,* 1989, *143*, 565.

12. L.-Y. Lian, J. C. Yang, M. J. Sutcliffe, J. P. Derrick, and G. C. K. Roberts, 1992, submitted for publication.

13. L-Y. Lian, J. C. Yang, J. P. Derrick, M. J. Sutcliffe, J. P. Murphy, C. R. Goward and T. Atkinson, *Biochemistry,* 1991, *30*, 5335.

14. K. Wüthrich, 'NMR of Proteins and Nucleic Acids', Wiley, New York, 1986.

15. P. Güntert, W. Braun and K. Wüthrich, *J. Mol. Biol.,* 1991, *217*, 517.

16. M. Nilges, G. M. Clore and A. M. Gronenborn, *FEBS Lett.,* 1988, *229*, 317.

17. S. Vijay-Kumar, C. E. Bugg and W. J. Cook, *J. Mol. Biol.,* 1987, *194*, 531.

18. A. M. Gronenborn, D. R. Filpula, N. Z. Essig, A. Achari, M. Whitlow, P. T. Wingfield and G. M. Clore, *Science,* 1991, *253*, 657.

19. G. Vriend and C. Sander, *Proteins,* 1991, *11*, 52.

20. J. P. Derrick and P. Dawson, unpublished work.

Experimental and Theoretical Studies of Enantioselective Receptors for Peptides

W. Clark Still, Shawn Erickson, Xuebao Wang, Ge Li, Alan Armstrong, Jong-In Hong, Sung Keon Namgoong and Ruiping Liu

DEPARTMENT OF CHEMISTRY, COLUMBIA UNIVERSITY, NEW YORK, NY 10027, USA

Biological receptors such as enzymes and antibodies are extraordinarily effective in binding their chosen substrates selectively. The typically high enantioselectivity of biological receptors is particularly striking and my group has sought to mimic this selectivity with synthetic receptors or host molecules. While the elementary forces controlling intermolecular associations (solvophobic effects, electrostatic and van der Waals interactions) seem to be well-understood, creating new synthetic receptors which work is still something of a black art. One key design strategy was defined by Don Cram and is the principle of preorganization: "the more highly hosts and guests are organized for binding and low solvation prior to their complexation, the more stable will be their complexes."[1] As we will show, the desirability for structural organization extends to the complex too. Thus selectivity among related guests is generally highest if host and guest bind together in only one way, i.e. if the host/guest complex is structurally homogeneous. In this paper, we will describe our studies of conformationally homogeneous receptors and the complexes they form with peptides or peptide-like guests.

Part 1. Primarily Experimental Studies

The first class of receptors we will discuss are nonmacrocyclic podand receptors for cations.[2] Such ionophores are related to the crown ethers but derive their conformational preorganization from 6-membered rings and a particular array of chiral centers and substituents. The glyme-ethers **1-3** are successively more and more preorganized. Whereas **1** and **2** have ~700 and 25 low energy conformations respectively, **3** has only one which is low in energy and capable of binding metal ions. It should provide an almost ideal template upon which to build chiral hosts.

Podand **3** is C_2 symmetric and chiral, and was synthesized in enantiomerically pure form from L-diethyl tartrate. As summarized in Table 1, **3** preferentially bound chiral ammonium hexafluorophosphates in CDCl$_3$ having the S configuration with enantiomeric excesses corresponding to ~0.5 kcal/mol.

Table 1. Enantioselective Binding of Podand Ionophore **3**.

Substrate	%ee
H$_3$N-CHMePh	42 (S)
H$_3$N-Ala-OMe	40 (S)
H$_3$N-Phe-OMe	36 (S)
H$_3$N-Val-OMe	34 (S)
H$_3$N-Met-OMe	36 (S)

The x-ray structure of **3**/(S)-α-phenethylammonium perchlorate is shown below in stereo:

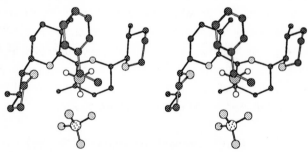

While a molecular mechanics conformational search of **3**/(S)-α-phenethyl-ammonium complex found as the global minimum a structure which was almost identical with the x-ray structure shown, the search found several other low energy conformations as well. These conformations were of two types. One had the phenyl in an axial-like orientation and was almost superimposable with the x-ray structure. The other was only a little higher in energy and essentially switched the positions of the guest phenyl and methyl as shown below:

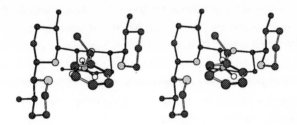

In solving crystal structures of other organoammonium complexes of **3** and related tetrahydrofuranoid ionophores, we subsequently found a complex which did show just such an equatorial-like orientation of the phenyl in the complex. These results provide a rationale for the low binding enantioselectivity we observe: while **3** is conformationally homogeneous, its complexes have multiple low energy forms.

Structural or conformational heterogeneity in hosts, guests or complexes is a major problem in molecular recognition. First, it generally fosters low selectivity in the binding of closely related guests, because different binding modes often have different selectivities. Second, it makes the structural basis of observed binding properties difficult to decipher. Finally, it provides a major impediment to quantitative modeling calculations because most simulation methods do not deal well with structures having multiple minima, especially if the minima are separated by significant energy barriers (see Part 2 of this paper).

We need to build recognition systems so that complexes as well as their unbound components will be conformationally homogeneous. Ideally we wish to construct binding sites which have complimentary steric and functional group interactions with the desired substrate in only one well-defined conformation and orientation relative to the receptor. How is this done? One answer, complimentarity based purely on steric effects, is difficult to attain because the often subtle differences between large (L), medium (M) and small (S) groups are difficult to distinguish with synthetically accessible receptors. In the figure below, such a situation is illustrated in the one-point associative binding mode shown. Here A⁻ and A⁺ are respectively receptor (R) and substrate functional groups which associate by strong intermolecular forces such as hydrogen bonds. While the specific A⁻/A⁺ association provides some structural stability to the complex by constraining the position of substituent A⁺, the substrate can still achieve a multiplicity of orientations relative to R by various rotations around A⁺ and around the A⁺-to-chiral center bond. Unless all significantly different orientations except for one are sterically forbidden, then the complex will be structurally heterogeneous.

An alternate two-point binding mode is intrinsically less flexible. Here, the complex is held together by two specific functional group associations, A+ with A- and B- with B+. In the host R, A- and B+ might represent hydrogen bond acceptor and donor groups respectively. With two-point binding, A+ and B- are both positionally fixed and the only intermolecular degree of freedom is rotation around the axis defined by A+ and B-. Such restrictions to flexibility in the complex should generally enhance binding selectivity.[3]

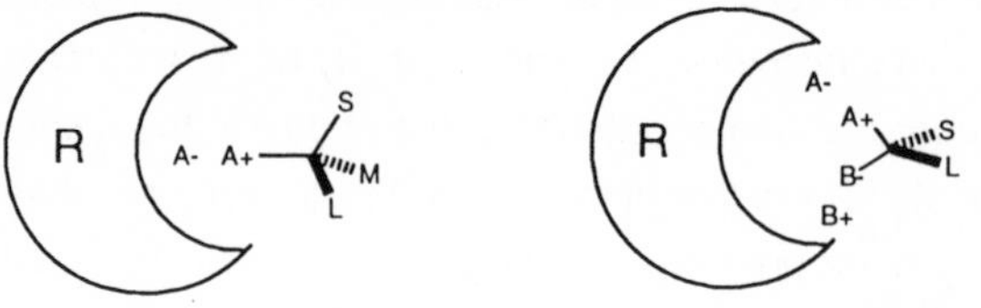

One Point   Two Point

The podand ionophores (e.g. **3**) discussed above correspond to the one-point binding model, but related ionophoric receptors which correspond to the two-point model can be designed. Such receptors are more heavily functionalized than **3** and several such hosts which we have prepared are shown below:

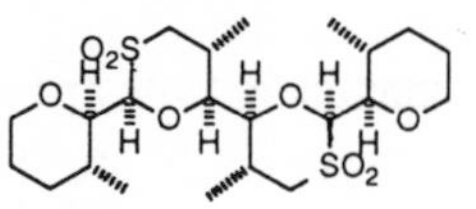

4

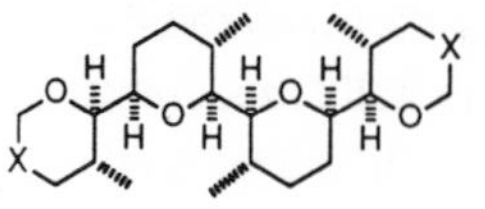

5a R = O
5b R = S

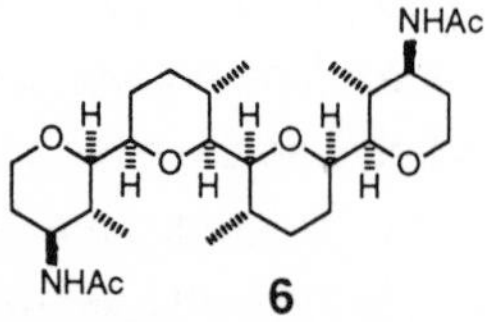

6

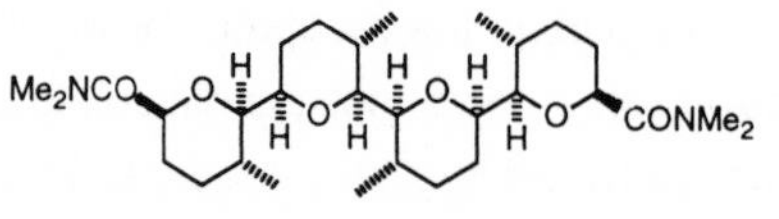

7

These receptors turn out to have significantly more interesting binding properties than does the simple podand **3**. The podand sulfone **4**, for example binds L-peptidic ammonium hexafluorophosphates with enantioselectivity as high as 80%.[4] While the primary interaction is still between the host and the guest ammonium ion, there appears to be a secondary electrostatic interaction between a sulfonyl S=O bond and the guest carboxyl group. An x-ray structure of the complex of **4** with L-proline methyl ester is shown below:

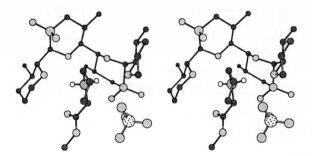

Essentially the same structure is found in the x-ray structure of **4** with L-proline N-methyl amide.

Podands **5a** and **5b** bind peptidic amides and esters differently. With amides, the best complexes form with peptides having the D configuration. With the corresponding esters, the L configuration is preferred. An x-ray structure of **5b** bound to the perchlorate salt of D-valine tert-butyl amide is shown below:

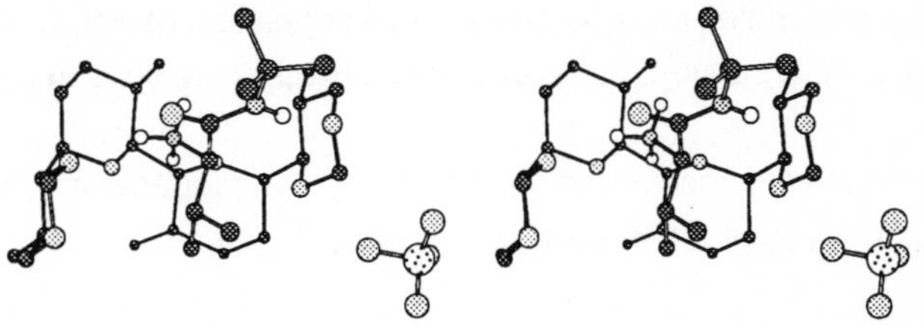

Note the hydrogen bond between the guest amide and one of the host acetal oxygens.

Podand **6** was designed to provide an acetamido hydrogen bond donor for secondary association with peptidic carboxyl groups. It enantioselectively binds peptidic esters (~60% ee) and amides (~80% ee or ~1.4 kcal/mol) having L configurations. While we have been unable to obtain experimental evidence for the proposed hydrogen bond, molecular simulations suggest it forms significantly more effectively with L than with D peptidic guests. Thus, we carried out 5000 ps of stochastic dynamics (which generates the Gibbs ensemble) at 300 °K and monitored the distances between the host donor and guest acceptor atoms (* in the diagrams below). The simulations of the L- and D-alanine complexes showed very different results as diagrammed below:

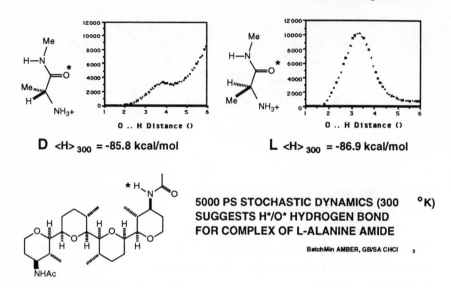

D <H> 300 = -85.8 kcal/mol   **L** <H> 300 = -86.9 kcal/mol

**5000 PS STOCHASTIC DYNAMICS (300 °K)
SUGGESTS H*/O* HYDROGEN BOND
FOR COMPLEX OF L-ALANINE AMIDE**

BatchMin AMBER, GB/SA CHCl$_3$

Whereas the D complex showed only a small population having O* and H* in proximity, the more stable L complex showed much more secondary hydrogen bond association.

The most enantioselective receptors we have prepared are the enantiomerically pure C_3-symmetric hosts shown below:[5]

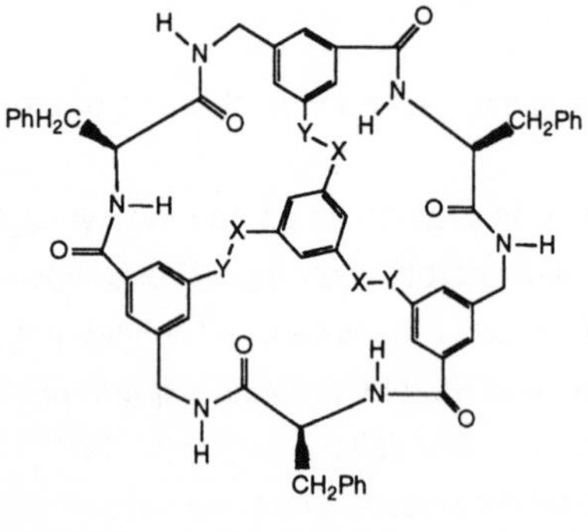

8a X = S, Y = CH$_2$
8b X = O, Y = CH$_2$
8c X = CH$_2$, Y = O

Extensive conformational searching suggests that, like **3-7** , these host molecules have a desirable structural property, i.e. conformationally homogeneity. They also form strong complexes with peptides in organic solvents with high enantioselection as shown in Table 2:

Table 2. Enantioselective Binding of C_3 Receptors in $CDCl_3$.

Substrate	8a %ee	8b %ee
N-Boc-Ala-NHMe	95 (L)	91 (L)
N-Boc-Val-NHMe	>99 (L)	99 (L)
N-Boc-Ser-NHMe	99 (L)	95 (L)
N-Boc-Thr-NHMe	>99 (L)	
N-Boc-His-NHMe	85 (L)	

While our C_3 receptors gave poor enantioselectivity with esters of simple amino acids having no sidechain functionality, we found that N-Boc methyl esters of glutamine, serine and threonine all gave enantioselectivity exceeding 80% ee for the L configuration.

We have been unable to crystallize **8** or its complexes; however, molecular modeling sheds light on the structure of the complexes involved. Again we used conformational searching to find the low energy states of the complex. This search was carried out with an internal coordinate Monte Carlo method[6] using a $CHCl_3$ solvation model and proceeded by varying the phi,psi angles of Boc-L-Ala-NHMe along with its position and orientation relative to the C_3 host. The lowest energy complex we found is shown below in stereo:

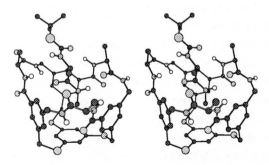

The complex is held together by three intermolecular hydrogen bonds with the guest C-terminal N-methyl group buried deeply within the cavity. Its position near the face of four aromatic rings suggests it should have an unusually high chemical shift in the ^1H nmr. Indeed, whereas the N-methyl resonates at 2.3 ppm in the unbound peptide, it is found at -0.8 ppm in the complex. Further support for the structure above comes from the observation of nOe signals between the guest N-methyl and the host trithiabenzene ring, and also from the fact that more heavily substituted amides (e.g. N-benzyl and tert-butyl) bind **8** only weakly.

Part 2. Primarily Theoretical Studies

One of our main goals is development of a computational method for screening new molecular designs for desirable properties. At this time, designing new catalysts, receptors, inhibitors, etc. is a rather empirical procedure in which the test of a new design is synthesis. Since it is usually easier to generate ideas than to test them by synthesis, we need ways to predict quantitatively molecular properties from molecular structure. While molecular modeling is the obvious choice for such predictions, it has not yet fulfilled its potential. And why not?

Part of the answer is that only recently have chemists had access to methods and hardware capable of simulating free energies of molecules in solution.[7] But even with our advances, there are areas which need further development if modeling is to become a useful predictive tool.

One area is the force field, the basic set of equations and parameters which relate energy to molecular geometry. As for the equations, we clearly need better treatments of hydrogen bonding and electrical polarization. We may need 5-body torsion-like terms to reproduce rotational energy profiles accurately. We also need to reevaluate many of the existing parameters for common functional groups or sets of proximate functionality. Fortunately, recent advances in quantum mechanics make it possible to obtain high quality energetic and structural data which is unavailable by experiment.

Another area in need of work is solvation, an effect which cannot be ignored if quantitative results are desired. Nearly all molecular mechanics calculations represent the solvent medium as a large, bounded set of explicit solvent molecules which move via Monte Carlo or dynamics through the Gibbs ensemble of states. While the explicit solvent model appears realistic, it is not perfect: the solvent molecules are typically rigid and electrically unpolarizable, the number of solvent molecules is typically only five times the number of atoms in the solute, different boundary conditions may give different final results, and the length of time a simulation runs may be insufficient to give adequate convergence. Such models leading to converged final energies may require more than 100 times the cpu time of the corresponding *in vacuo* calculations. An appealing alternative to an explicit solvent model is a continuum model which represents solvent as a continuous medium surrounding the solute and having the average properties of solvent. While such continuum solvent models say nothing about the structure of solvent, they do a surprisingly good job at reproducing the thermodynamic properties of solutes.[8,9]

For many years, continuum models either represented the solute as a simple ellipsoid or used numerical methods which were inappropriate for use in molecular mechanics or dynamics. More recently, our group developed a differentiable, analytical treatment of solvation based on the generalized Born equation which has proven a promising alternative to explicit solvent models.

The model we developed is called the GB/SA (generalized Born/surface area) continuum solvation model[9] and it relates molecular surface areas, atomic positions, charges and Born radii to solvation energies ($E_{solvation}$) as follows:

$$E_{solvation} = E_{cav} + E_{vdW} + E_{pol} \qquad \text{Eq 1}$$

$$E_{cav} + E_{vdW} = \Sigma \, \sigma_i \, A_i \qquad \text{Eq 2}$$

$$E_{pol} = -166 \, (1-1/\varepsilon) \, \Sigma_i \Sigma_j \, q_i q_j / (r_{ij}^2 + \alpha_{ij}^2 e^{**}(-r_{ij}^2/4\alpha_{ij}^2))^{0.5} \qquad \text{Eq 3}$$

In Eq 2, σ_i is an atomic solvation parameter and A_i is the solvent accessible surface area of atom i. We determine σ empirically from experimental solvation energies of simple hydrocarbons, molecules for which E_{pol} should be relatively insignificant. In Eq 3, ε is the bulk dielectric constant of the solvent, Σ_i and Σ_j define a double sum running over all pairs (i,j) or atoms in the molecule, q_i and q_j are atomic charges on atoms i and j, r_{ij} is the separation between atoms i and j, and α_{ij} is a mean Born radius of the atom i j pair (we use $\alpha_{ij} = (\alpha_i \alpha_j)^{0.5}$). For a given solvent, the model has only two empirically determined parameters (σ above and a dielectric offset distance used in computing α). The solvation equations are also rapid to evaluate - a GB/SA molecular mechanics or dynamics calculation is only ~3-fold slower than the corresponding *in vacuo* calculation.

We tested the calculation against experimental solvation free energies in water for small organic molecules whose parameters (most significantly partial charges) were reported in the literature. The tests included a variety of solutes including alcohols, esters, acids, amides, ketones, simple peptides, aromatics and ions such as ammonium and acetate. The correlation between experiment and calculation was quite high ($R^2 = 0.96$) as shown on the following page for neutrals.

Since developing the model, we have optimized GB/SA parameter sets for water and chloroform, and used the model extensively in our molecular recognition modeling studies. These studies include conformational searching in solution and free energy perturbation calculations of binding selectivity.

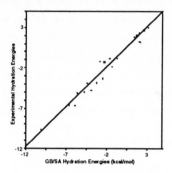

One of the host/guest systems we have studied using free energy perturbation is our C_3 receptor **8**. Starting from the global minimum of the Boc-L-Ala-NHMe complex shown previously, we mutated the L-alanine into D-alanine by transforming the alanine α-H and α-Me to α-Me and α-H respectively. We used our GB/SA model for chloroform in MacroModel and carried out a total of 2000 ps of stochastic dynamics in 20 stages for the mutation. The results of our free energy perturbation calculation are summarized in the graph below:

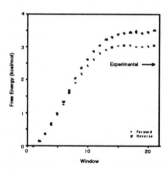

Thus in the mutation of bound L-Ala into D-Ala, we calculate the free energy to increase by ~3.2 kcal/mol in comparison with an experimental increase of 2.2 kcal/mol. While the calculation is qualitatively in accord with experiment, the error of 1 kcal/mol is nontrivial. We have carried out many such mutations of chiral guests in chiral binding sites and find similar results as those shown above: our calculations are in qualitative agreement with experiment but errors as high as 1.5 kcal/mol are not uncommon.

The errors in our calculations have several sources. One source is the quality of the force field itself. There is recent evidence that standard force field parameters do not accurately reproduce the *in vacuo* behavior of some of the peptidic substructures in **8** and its guests.[10] Consequently we are testing our force field parameterization against high level *ab initio* results and reparameterizing where necessary.

Another source of error is inadequate sampling in the dynamics simulations. Although a total simulation of 2000 ps is lengthy by contemporary standards, we find that our results vary as much as 1 kcal/mol with longer simulations or with different initial conditions.[11] Clearly, we are not sampling the various significantly populated conformations or configurations of the system well enough to obtain converged results. While significantly longer simulations could be used, we think there is a better way to sample the populated states of the system. To this end we have recently devised a mixed mode dynamics/Monte Carlo procedure which uses dynamics to investigate local conformational space and internal coordinate Monte Carlo (MC) make transitions between the various minima. Since we make very large MC internal coordinate changes (e.g. torsion angles ±180°), barriers between minima are inconsequential. The procedure we use operates by attempting an MC step every 1-10 (stochastic) dynamics timesteps. Most MC steps fail and the dynamics simulation continues. Whenever a MC step passes the Metropolis test, the new geometry is accepted, velocities are resampled and the dynamics is continued with the new geometry. This method works well only with a continuum solvent because explicit solvent molecules interfere with large MC movements. We are now applying this method to our free energy perturbations.

It looks as if the use of continuum solvent and sampling methods which allow large changes in internal coordinates may yield a big win for simulations directed toward calculating thermodynamic properties. In our own systems, these methods yield results which are comparable to traditional simulation procedures requiring 3-4 orders of magnitude more computer time. We appear to be approaching a level of accuracy and efficiency that will soon allow us to use molecular modeling as a practical, quantitative design tool.

Notes and References.

1. D.J. Cram, *Science*, **240**, 760 (1988).
2. T. Iimori and W.C. Still, *J. Am. Chem. Soc.*, **111**, 3439 (1989); T. Iimori, S.D. Erickson, A.L. Rheingold and W.C. Still, *Tetrahedron Lett.*, **30**, 6947 (1989); S.D. Erickson and W.C. Still, *Tetrahedron Lett.*, **31**, 4253 (1990); X. Wang, S.D. Erickson, T. Ilmori and W.C. Still, *J. Am. Chem. Soc.*, in press (1992).
3. Similar conclusions have been reached by others: W.H. Pirkle and T.C. Pochapsky, *Chem. Rev.*, **89**, 347 (1989); P.P. Castro and F. Diederich, *Tetrahedron Lett.*, **32**, 6277 (1991).
4. G. Li and W.C. Still, *J. Org. Chem.*, **56**, 6964 (1991).
5. J.-I. Hong, S.K. Namgoong, A. Bernardi and W.C. Still, *J. Am. Chem. Soc.*, **113**, 5111 (1991).
6. G. Chang, W.C. Guida and W.C. Still, *J. Am. Chem. Soc.*, **111**, 4379 (1989).
7. General review: W.L. Jorgensen, *Chemtracts - Organic Chemistry*, **4**, 91 (1991).
8. A. Warshel and S.T. Russel, *Q. Rev. Biophys.*, **91**, 4105, 4109, 4118 (1987); M. Gilson and B. Honig, *Proteins*, **4**, 7 (1988).
9. W.C. Still, A. Tempczyk, R.C. Hawley and T. Hendrickson, *J. Am. Chem. Soc.*, **112**, 6127 (1990); C.J. Cramer and D.G. Truhlar, *J. Am. Chem. Soc.*, **113**, 8305 (1991).
10. E.g. H.-J. Bohm and S. Brode, *J. Am. Chem. Soc.*, **113**, 7129 (1991)
11. E.g. M.J. Mitchell and J.A. McCammon, *J. Comput. Chem.*, **12**, 271 (1991); W.F. van Gunsteren and A.E. Mark, *Eur. J. Biochem.*, in press (1992).

Enabling Methodology: The Synthetic Chemist's Contribution to Molecular Recognition

Steven V. Ley and Lam Lung Yeung

DEPARTMENT OF CHEMISTRY, IMPERIAL COLLEGE OF SCIENCE, TECHNOLOGY AND MEDICINE, SOUTH KENSINGTON, LONDON, SW7 2AY, UK

INTRODUCTION

Organic Synthesis plays a vital and underpinning role in our understanding of molecular recognition processes. As our knowledge in this area advances so do the demands placed upon current methods for synthesis. There is a constant need therefore to discover new strategies and methods which will match the strides being made in molecular biology, crystallography and molecular modelling techniques. For these reasons we have investigated a new approach towards important biologically active phosphoinositols. There is currently considerable interest in molecules of this type since they are implicated in a vast range of molecular recognition processes. For example, D-*myo*-inositol-(1,4,5)-trisphosphate (Ins(1,4,5)P$_3$) and the related phosphatidylinositol-(4,5)-bisphosphate (PtdIns (4,5)P$_2$), are important in fundamental cell-signal transduction mechanisms,[1,2] particularly in calcium mobilisation[3,4] within the cell. Many receptor controlled mechanisms are known to involve Ins(1,4,5)P$_3$ and its involvement in the phosphoinositide cycle (PI cycle).[2] Other phosphoinositol species stimulating interest and synthetic studies are the putative insulin mediators (PIMs). These compounds are thought to be derived from the binding of insulin to plasma membrane receptors and can mimic biochemical activities of the hormone on certain target enzymes.[5,6] They can also be produced by the combined actions of protease and phospholipase C on glycosyl phosphatidylinositol (GPI).[7] This novel class of GPI has recently been characterised from the *Trypanosoma* variant surface glycoproteins (VSGs)[8,9] and have been identified as membrance anchors for a variety of cell surface proteins.[10] Their importance in cell adhesion, antigen activity and recognition[10,11] is the focus of many research programmes (Scheme 1).

We have sought to develop a novel approach to the preparation of the central inositol unit common to all these compounds whereby we can control the hydroxyl group substitution patterns in a regio- and stereo-selective fashion. In this way we might be able to achieve the preparation of all members in the series, and be able to supply large quantities of material and novel analogues for biological studies. Previous routes to these compounds have largely relied upon multistep modification of *myo*-inositol itself.[12] While these reactions are successful they are limited in the development of the required functionality. Our approach utilizes a biotransformation

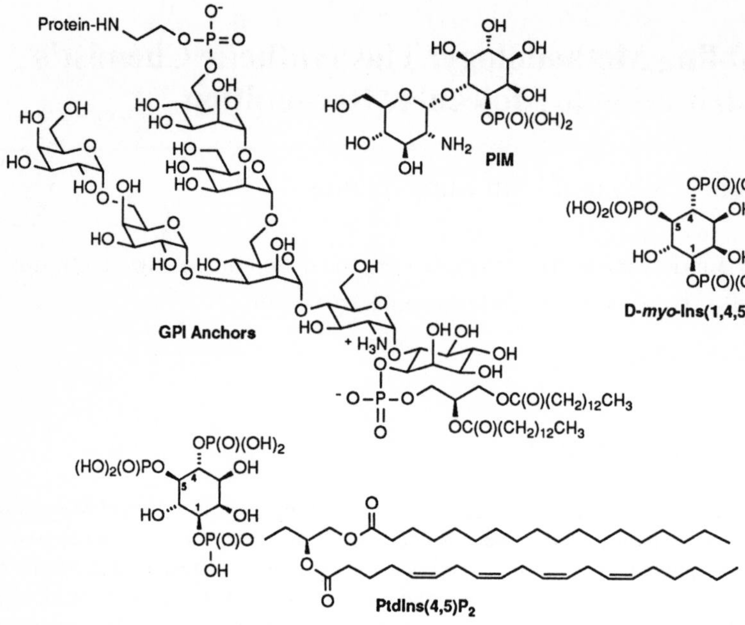

Scheme 1

to set up the initial oxidation pattern using arenes as precursors with the microbial oxidant *Pseudomonas putida*[13] (Scheme 2).

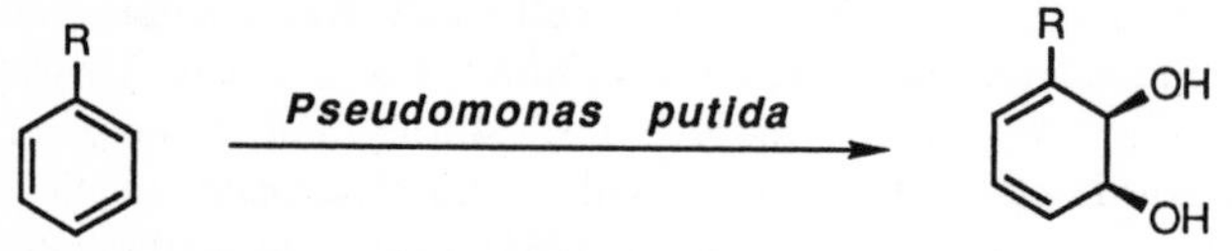

Scheme 2

We believe this to be a strategically important reaction since it cannot be achieved with conventional synthetic reagents.[14] Since we first pointed out the significance of this process in providing novel starting materials for synthesis from arenes, many other groups worldwide have adopted these methods.[15] The versatility of these cyclohexadiene diols for further transformation is best illustrated by the generalities expressed in Scheme 3.

SYNTHESES

It is obvious that these cyclohexadiene diols should provide a superb starting point for inositol/cyclitol synthesis. Indeed we have shown that the product derived by oxidation of benzene may be readily converted to many biologically active compounds. For example the hypoglycemic, antidiabetic natural product pinitol[16] was obtained in 40% overall yield in just five steps (Scheme 4). This represents a very efficient process for setting up the six contiguous asymmetric centres in a

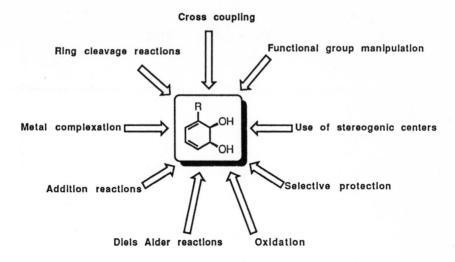

Scheme 3

cyclic array. By slight modification of the pathway we can also prepare the enantiomerically pure compounds (+) and (-) pinitol.[17]

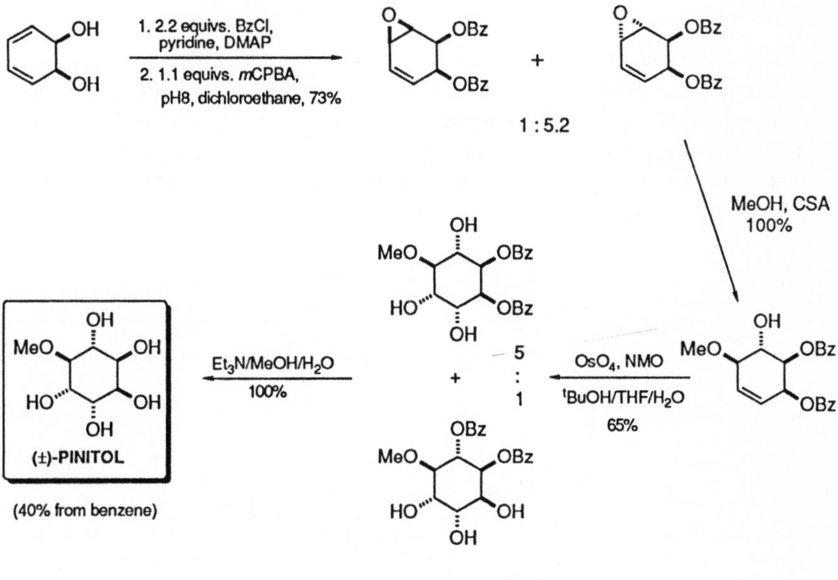

Scheme 4

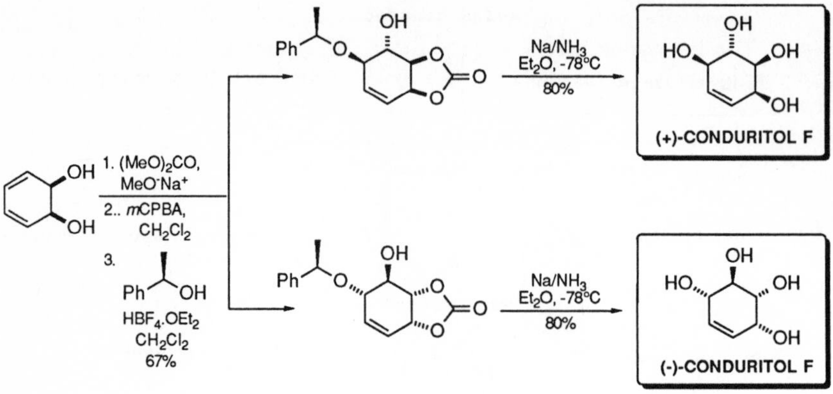

Scheme 5

In a similar way we have also achieved the preparation of (+) and (-) conduritol F[18] (Scheme 5). (+)-Conduritol F is a component of almost all green plants. Related polyhydroxylic systems are also useful glycosidase inhibitors.[19] The above synthesis is extremely short, only five steps from benzene, and affords enantiomerically pure compounds by making use of the diastereomer separation arising by epoxide ring opening with R-(+)-*sec*-phenethyl alcohol. The common building blocks which are generated in these sequences are also useful in many of the later syntheses. This was an important feature of the design of these synthetic schemes.

We have also recently used one of these building blocks in the preparation of carba-α-D-glucopyranose, an important inhibitor of glucokinase and glucose stimulated insulin release[20] (Scheme 6). The key steps in this synthesis are the regio- and stereo-selective ring opening of the epoxide by an acetylide anion to introduce the required side-chain, and the reduction of the intermediate triflate with super-hydride (LiEt3BH) to establish the methylene unit.[21]

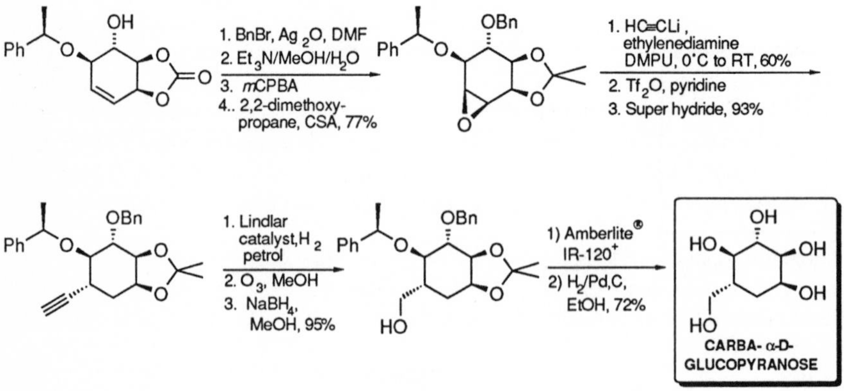

Scheme 6

We next turned our attention to the preparation of the key members of the PI cycle. The important synthetic issues which arise are; how to obtain suitably protected inositols; how to effect efficient and possibly selective phosphorylation and how to obtain material in enantiomerically pure form. Furthermore, we require mild deprotection conditions to avoid phosphate scrambling[22] and eventually prepare novel analogues or species which contain very labile acyclic glyceride side-chains such as arachidonic esters. In the first instance we devised a route to D(-)-Ins(1,4,5)P$_3$[23] (Scheme 7).

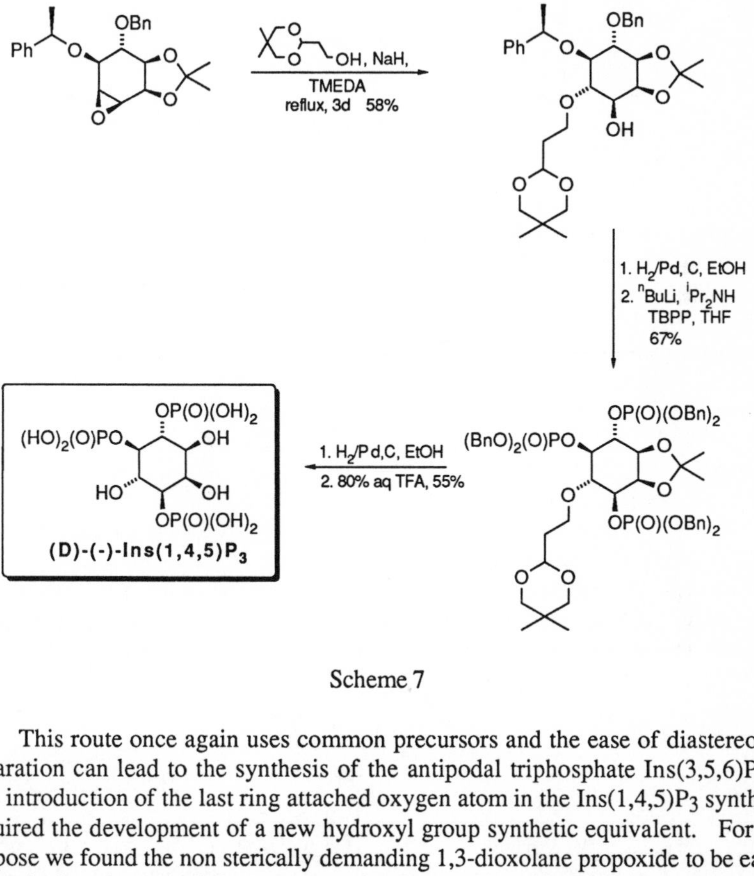

Scheme 7

This route once again uses common precursors and the ease of diastereomer separation can lead to the synthesis of the antipodal triphosphate Ins(3,5,6)P$_3$.[23] The introduction of the last ring attached oxygen atom in the Ins(1,4,5)P$_3$ synthesis required the development of a new hydroxyl group synthetic equivalent. For this purpose we found the non sterically demanding 1,3-dioxolane propoxide to be easily unmasked at the end of the sequence simultaneously with acetonide deprotection (Scheme 7). As we also wished to prepare novel analogues for these common precursors, we synthesised 6-Me-Ins(1,4,5)P$_3$[24] (Scheme 8) which was a full agonist for calcium release but several orders less potent than Ins(1,4,5)P$_3$.[25] Three other analogues with various C-6 substituents were also prepared by standard methods (Scheme 9). These reactions have been described in more detail elsewhere.[23]

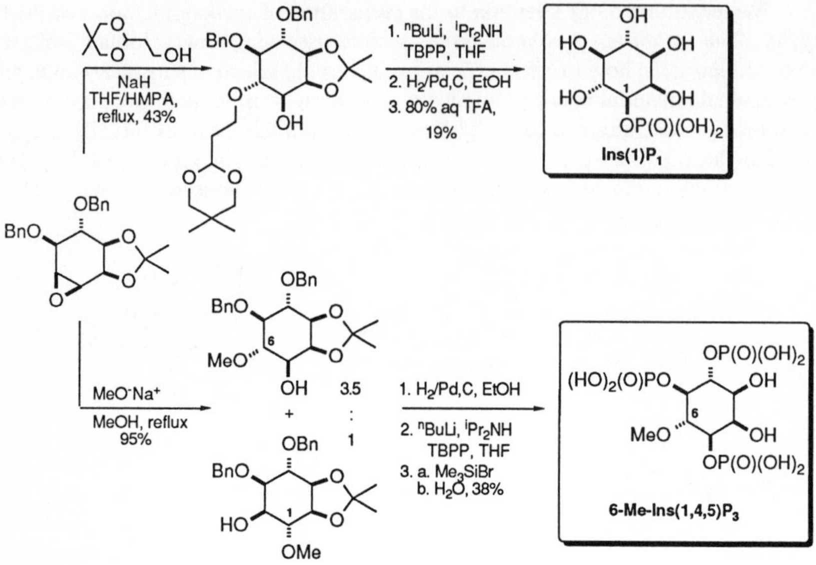

Scheme 8

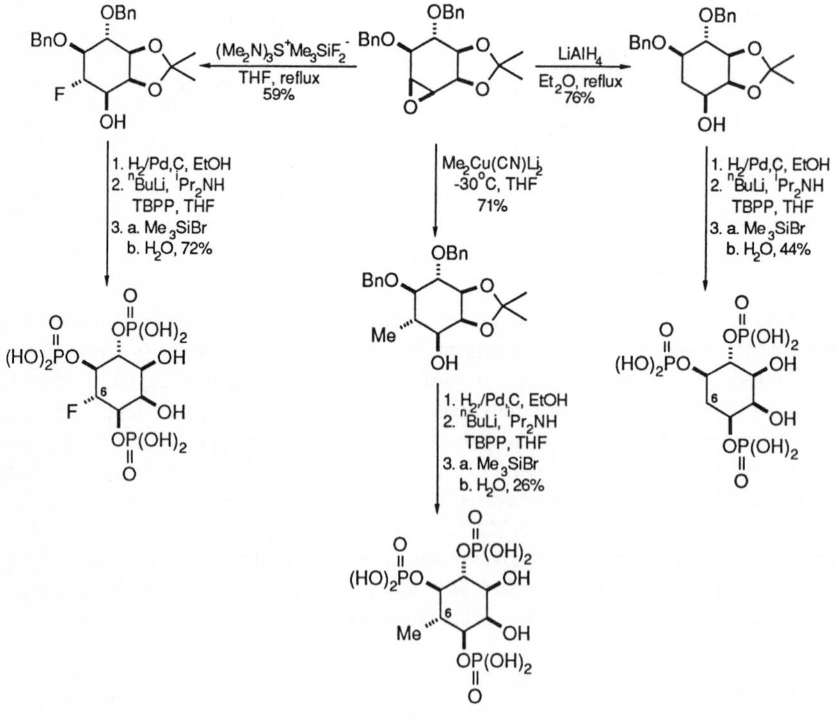

Scheme 9

We have also made significant progress on the synthesis of Ptd Ins(4,5)P$_2$ (PIP$_2$). Both the required selectively protected inositol with a free hydroxyl group at C-1 and protected phosphates at C-4 and C-5 (Scheme 10) and the necessary glyceride side chain for coupling at C1 are now available in large quantities. We are presently examining these reactions to complete the first synthesis of this biologically important compound.

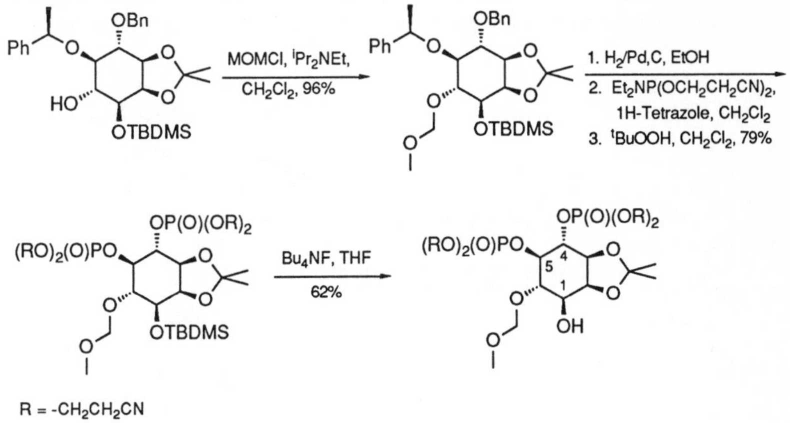

Scheme 10

In order to achieve the synthesis of the most complex inositol target molecules, namely the GPI anchors, we need to establish routes to appropriately 1,6-differentiated inositol precursors. Since the same substitution patterns also occur in the related putative insulin mediators (PIMs), we have studied these systems as a prelude for our GPI work. Finally, we would like to report the preparation of 6-*O*-(2-amino-2-deoxy-α-D-glucopyranosyl)-1-*O*-dihydrogen-phosphonyl-D-*myo*-inositol (PIM). This was achieved again using a previously synthesised common precursor epoxide (Scheme 11).

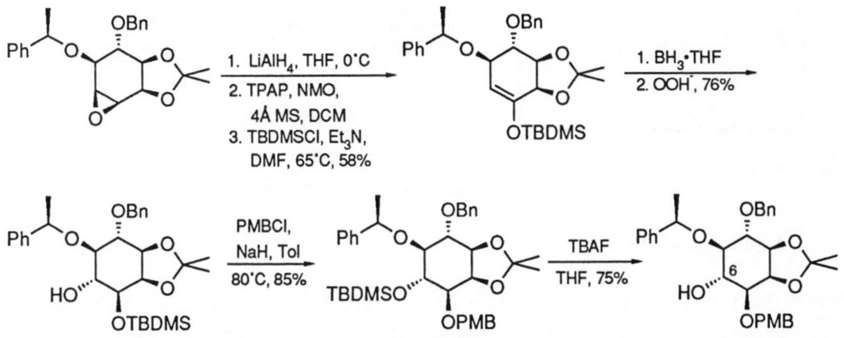

Scheme 11

It should be noted that the first steps of this scheme afford the same starting material necessary for the PIP$_2$ synthesis discussed earlier. The last steps of the preparation of the PIM make use of the Schmidt coupling of an azido-glucosamine derivative, producing a 2.5:1 α:β anomer ratio (Scheme 12). The protecting groups used in this sequence were chosen such that they could be removed under the reduction conditions used for conversion of the azide group to the required amine substituent. The final acetonide was readily removed by rapid treatment with aqueous acetic acid. These reactions not only afford the PIM compound, but also set the scene for our route to the GPI anchor which we will describe at a later date.

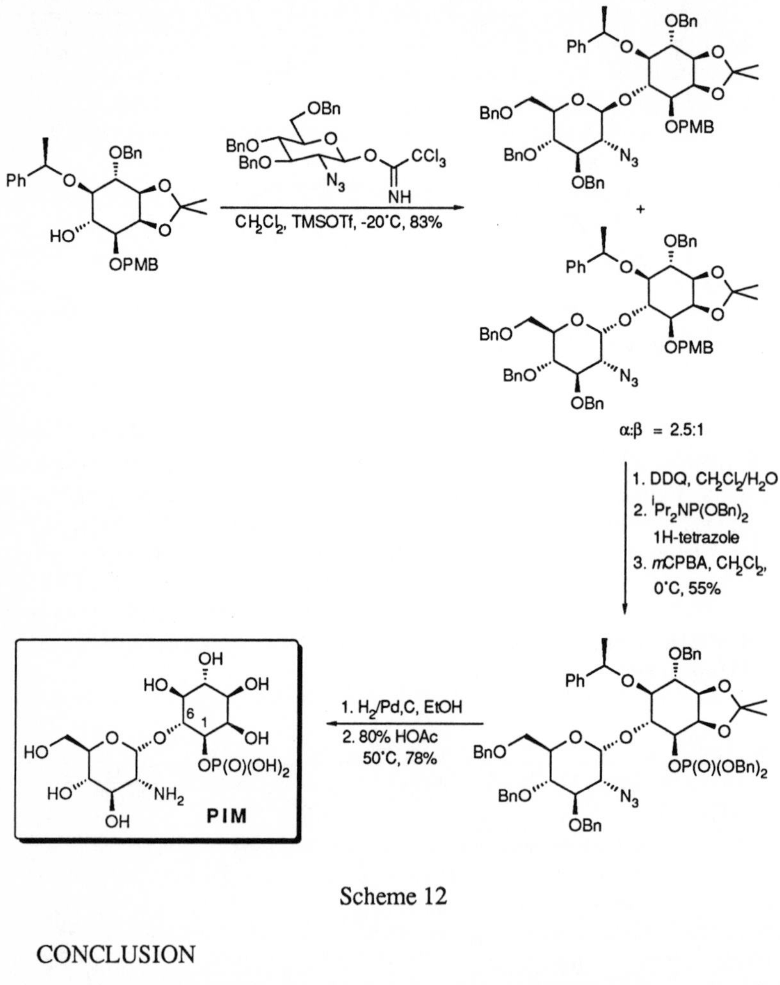

Scheme 12

CONCLUSION

We believe we have developed a new strategically important process which leads to the preparation of biologically important inositol phosphates and structural analogues. We anticipate many further applications of these systems in organic synthesis, leading to interesting compounds for molecular recognition studies.

ACKNOWLEDGMENTS

We thank the Croucher Foundation for a research scholarship to YLL, S. C. Taylor for generous samples of *cis*-1,2-dihydroxycyclohexa-3,5-diene and Glaxo Group Research for financial support.

REFERENCES

1. M.J. Berridge and R.F. Irvine, *Nature*, **1984**, *312*, 315.
2. M.J. Berridge and R.F. Irvine, *Nature*, **1989**, *341*, 197.
3. R.H. Michell, *Biochim. Biophys. Acta.*, **1975**, *415*, 81.
4. M.J. Berridge, *A. Rev. Biochem.*, **1987**, *56*, 159.
5. A. Saltiel and P. Cautrecasas, *Proc. Natl. Acad. Sci., USA.*, **1986**, *83*, 5793.
6. J. M. Mato, K. L. Kelly, A. Abler and L. Janett, *J. Biol. Chem.*, **1987**, *233*, 2131.
7. M.P. Czech, J.K. Klarlund, K.A. Yagaloff, A.P. Bradford and R.F. Lewis, *J. Bio. Chem.*, **1988**, *263*, 11017.
8. M.A.J. Ferguson, S.W. Homans, R.A. Dwek and T.W. Rademacher, *Science*, **1988**, *239*, 753.
9. B. Schmitz, R.A. Klein, I.A. Duncan, H. Egge, J. Gunawan and J. Peter-Kalalinic, *Biochem. Biophys. Res. Commun.*, **1987**, *146*, 1055.
10. M.A.J. Ferguson and A.F. Williams, *Ann. Rev. Biochem.*, **1988**, *57*, 285.
11. M.G. Low, *Biochem. J.*, **1987**, *244*, 1.
12. B.V.L. Potter, *Natural Product Reports*, **1990**, *7*, 1; D.C. Billington, *Chem. Soc. Rev.*, **1989**, *18*, 83.
13. D.T. Gibson, J.R. Koch and R.E. Kallio, *Biochemistry*, **1968**, *7*, 2653.
14. M. Nakajima, I. Tomida and S. Takei, *Chem. Ber.*, **1959**, *92*, 163.
15. D.A. Widdowson and D.W. Ribbons, *Janssen Chimica Acta.*, **1990**, *8*, 3; S.M. Brown, 'Organic Synthesis: Theory and Applications'; T. Hudlicky, Ed.; JAI Press Inc.: Greenwich, Connecticut, **1992**; Vol. 2, in press.
16. C.R. Narayanan, D.D. Joshi, A.M. Miyumdar and V.V. Dhekne, *Curr. Sci.*, **1987**, *56*, 139.
17. S.V. Ley and F. Sternfeld, *Tetrahedron*, **1989**, *45*, 3463.
18. S.V. Ley and A. J. Redgrave, *Synlett*, **1990**, 393.
19. G. Legler and E. Bause, *Carbohyd. Res.*, **1973**, *28*, 45.
20. I. Miwa, H. Hara, H. Okuda, T. Suami and S. Ogawa, *Biochem. Int.*, **1985**, *11*, 809.
21. S.V. Ley and L.L. Yeung, *Synlett*, **1992**, 291.
22. D.J. Cosgrave, "Inositol Phosphates, their Chemistry, Biochemistry and Physiology", Elsvier, Amsterdam, 1980.
23. S.V. Ley, M. Parra, A.J. Redgrave and F. Sternfeld, *Tetrahedron*, **1990**, *46*, 4995.
24. S.V. Ley and F. Sternfeld, *Tet. Lett.*, **1988**, *29*, 5305.
25. S.T. Safrany, R.J.H. Wojcikiewicz, J. Strupish, S.R. Nahorski, D. Dubreuil, J. Cleophax, S.D. Gero and B.V.L. Potter, *FEBS Lett.* **1991**, *278*, 252.

Subject Index